The Sault Ste. Marie Canal

A Chapter in the History of Great Lakes Transport

Brian S. Osborne and Donald Swainson

Studies in Archaeology
Architecture and History

National Historic Parks and Sites Branch
Parks Canada
Environment Canada

©Minister of Supply and Services Canada 1986.

Available in Canada through authorized bookstore agents and other bookstores, or by mail from the Canadian Government Publishing Centre, Supply and Services Canada, Hull, Quebec, Canada K1A 0S9.

La traduction française s'intitule **Le canal de Sault-Sainte-Marie: Un chapitre de l'histoire des transport sur les Grands Lacs** (n⁰ de catalogue R61-2/9-32F). En vente au Canada par l'entremise de nos agents libraires agréés et autres librairies, ou par la poste au Centre d'édition du gouvernement du Canada, Approvisionnements et Services Canada, Hull, Québec, Canada K1A 0S9.

Price Canada: $7.50
Price other countries: $9.00
Price subject to change without notice.

Catalogue No.: R61-2/9-32E
ISBN: 0-660-12066-6
ISSN: 0821-1027

Published under the authority
of the Minister of the Environment,
Ottawa, 1986.

Editing and layout: Paula Irving
Cover Design: Jean Brathwaite and Suzanne Adam-Filion

The opinions expressed in this report are those of the authors and not necessarily those of Environment Canada.

Parks Canada publishes the results of its research in archaeology, architecture and history. A list of titles is available from Research Publications, Parks Canada, 1600 Liverpool Court, Ottawa, Ontario, K1A 1G2.

Cover: The Sault Ste. Marie ship canal.

CONTENTS

4 Acknowledgements
5 Introduction

I The First Contact: The "Bawatig" and the Fur Companies
7 The "Bawatig"
12 The Fur Traders
20 The Close of the First Contact

II Bypassing the Rapids: The Case for the Canals at the Sault
22 The "Soo" at Mid-Century
25 Early Developments
36 The Sault as a Junction

III Construction of the Canadian Canal
38 The Government Committed
41 The Long Wait
43 The Final Decision to Build
46 Awarding the Contracts
53 Progress of Construction
73 Completion and Opening

IV Operating the Canal
75 Operating the New System
80 The Establishment at the Sault
85 Collisions, Groundings and Blockages
93 Canal Traffic
108 The "Soo" as a System

V Epilogue: The Rapids, the Canal and the Town
109 From "Suburbs" to Municipality
112 Economic Links
117 Social Links
119 Protecting the Lock
123 Continuity and Heritage

125 Endnotes
136 Glossary
137 Bibliography

Submitted for publication in 1984 by Professor Brian S. Osborne and Professor Donald Swainson, Queen's University, Kingston.

ACKNOWLEDGEMENTS

This study has benefitted from the contributions of several Parks Canada staff at Cornwall and Sault Ste. Marie. Dennis Carter-Edwards and John Witham at the Cornwall office provided invaluable assistance. Norman Ruttan contributed much to the study in identifying local resources and providing an "in the trenches" awareness of matters of concern to the staff at the Sault. His colleague Roger Draycott made a valuable contribution to this project through his photographic services.

Competent professional archival assistance was offered willingly by Glen Wright at the Public Archives of Canada and John Mezaks at the Archives of Ontario. The cooperation of the Hudson's Bay Company in allowing us access to their materials relating to the Sault must also be acknowledged. Other depositories consulted include the Bayliss Library, Sault St. Marie, Michigan; the collection of the Sault St. Marie Historical Society, the Armories, Sault Ste. Marie, Ontario; and the Canadian Pacific Railways Archives. We gratefully acknowledge the assistance provided by the staffs at these depositories.

We also thank research assistants Carolyn Abrahams, John Belec, Francis Heath, Lois Edmonds and Bill Teatero; Ross Hough and George Innes for cartography and photography services, respectively; and special thanks to Joan Knox.

INTRODUCTION

The significance of the St. Lawrence—Great Lakes waterway in Canadian history has been well argued by several scholars. As a transport route and corridor of movement it has played an important role in the economic, political and social development of the nation. As such, in several periods of Canadian history it has moved people and commodities across the length of the country, the commodities and modes of transport changing as new regions were opened, new products developed and new technologies introduced. French canoes in the fur trade, British batteaux demarcating the frontier of the colony, Canadian-American schooners and steamships plying their trade in the agricultural and industrial staples of a growing commerce, all used this dominant continental drainage system.

Extensive as it was, the potential of the St. Lawrence—Great Lakes system was limited in its original form by the considerable rapids of the St. Lawrence, the magnificent obstacle of the Niagara Falls between Lakes Ontario and Erie, and the less grand but no less restricting rapids between Lakes Huron and Superior at Sault Ste. Marie. As long as the staples were portable and of a high value to bulk ratio, canoes, portages and strong men bypassed such obstacles. As the cargoes increased in both volume and diversity, the need to accommodate the through transport of carriers of larger capacity became essential. The initially modest locks around the rapids of the St. Lawrence, the ambitious if shortsighted Rideau project bypassing the St. Lawrence entirely and the inspiring works required to overcome the obstacle of the Niagara escarpment were all early, if limited, attempts at improving the system. By the mid-19th century, lake steamers could travel from the lakes, along the St. Lawrence, to tidewater at Montreal.

While improvements continued at these sites throughout the century, interest shifted west to the last major obstacle to shipping — the rapids separating Lakes Superior and Huron. The development of a Canadian canal at Sault Ste. Marie was the Macdonald government's response to economic and nationalist arguments by entrepreneurs interested in exploiting the rich natural resources of the Lake Superior region and providing an all-Canadian route for the passage of grain from the rapidly developing Canadian West. But at a time of parallel canal developments on the American side of the river, the Sault Ste. Marie Canal's initial advantage in depth and length was surpassed by its rival and the canal was relegated to a secondary role in terms of tonnage and lockages. It did, however, provide at the time of its completion in 1895 the final link in the all-Canadian chain of improvements stretching from Lake Superior to the Atlantic coast. The Canadian canal served well the passenger ships and small freighters that plied the Great Lakes and was essential to the industrial growth of the adjacent town. Finally, the canal was important for its pioneering adaptation of hydroelectric power to canal operations, putting Canada in the forefront of innovative canal technology at the turn of the century. This fact was recognized when the canal operations were transferred to Parks Canada in 1979.

This study tells the story of the several periods of interaction with the rapids at the Sault. At each period in the contact, the presence of the rapids was the rationale for the location there of the successive groups, each with different principles of social and economic organization. Despite several developments and modifications, the rapids still dominate the Sault Ste. Marie landscape.

1 Sault Ste. Marie from the Canadian side by J.H. Caddy. (Courtesy of the Royal Ontario Museum, Toronto, Canada)

I THE FIRST CONTACT: THE "BAWATIG" AND THE FUR COMPANIES

For much of its history, Sault Ste. Marie has been considered a transit zone, a corridor of movement, a conduit for traffic moving east and west along the Laurentian waterway. This was certainly as true for the Indians of the Great Lakes region as it was for the later fur traders and the subsequent political and economic interests associated with the national drive to the West.

For several centuries, however, there was an additional perspective. For both the indigenous population and the later fur trade interests, the Sault was an important focal point not only because of communication considerations but also because of the fishery at the rapids. To both groups, the convenient junction of routes was also an important resource.

The "Bawatig"

In the regional geography of the Indian world, the Sault functioned as transport junction, resource and an equally important ceremonial centre.[1] Indeed, from the French period contact, the Algonkan peoples of this area were referred to as Saulteur, and those Saulteur associated with the "Bawatig" (the Algonkan word for rapids) were called by Claude Dablon, a 17th-century missionary, the "Pahouitingwach Irini" or "Baouitchtigouian."[2] Later simplifications modified this nomenclature to "Batchawana," the group associated by both residence and specialized skills with the Sault rapids.

For other Algonkan groups, adaptation to the Canadian Shield required seasonal migration which dispersed many groups in small units throughout the interior forests during the winter, their main sustenance being the yield from hunting. Spring migrations of fish and fowl attracted them to estuarine locations along the lakes, where there were often sugar maples. The gathering of wild rice in the late summer, and the fall runs of certain fish species supported other concentrations of population before the dispersions to the winter hunting grounds. Sturgeon and whitefish predominated, but pike, lake trout, pickerel and other smaller species were included in the diet. The preservation of surplus fish by smoking and drying provided both supplements for the winter diet and also commodities for trade with other groups such as corn from the Hurons.[3] This pattern continued through the summer but it was the run of fish from fall until the onset of winter that constituted the most productive fishery. During this season the white fish moved in from the deep waters of the lake to the shallow waters of the river estuaries to spawn. At such times, catches of several hundred a night were not uncommon.[4]

The Sault rapids was one of these sites of seasonal concentration of population because of its prodigious yields of whitefish. Although

Étienne Brûlé and Jean Nicollet probably visited the Sault on their explorations between 1621 and 1623 and in 1634, respectively, and Indian traders had spread knowledge of the fishery there to more easterly locations, it was not until 1641 that the Jesuits Jogues and Raymbault provided the first commentary on the community at the rapids. Arriving in the fall of 1641, they were met by some 2000 people of the Sault, most of whom had travelled there from distant areas.[5] Father Dablon described the native dependence on the whitefish, the "Atticameg," and commented with awe that the "convenience of having fish in such quantities that one has only to go and draw them out of the water, attracts the surrounding nations to the spot during the summer." These "wanderers," as he called them, were said to include the Noquet from south of Lake Superior, the Outchibous, the Marameg, the Achilgouiane, the Amicourses, the Mississaugas, together with some eight other groups from as far away as the "North Sea" and Lake Winnipeg.[6]

Although these groups integrated the whitefish at the rapids into their seasonal itinerant cycle, apparently the "Pahouitingwach Irini," or Saulteur, lived there permanently. Not only did a productive resource base allow such permanence, but the permanence allowed the development of the specialized skills required to make optimal use of the resource. Others may have availed themselves of the seasonal runs of fish along the shores of the rapids but it was the Saulteur who were able to exploit the treacherous waters of the rapids. Thus, although the Jesuits reported a seasonal concentration of population at the Sault amounting to some 2000 Indians, only 200 Saulteur were permanent residents there. Though limited in numbers, their specialized function as harvesters of the yield of the rapids ensured for them a particular claim to residence at the Sault. This was the conclusion of Father Hennepin who early recognized the association between the distinctiveness of the Saulteur and the special skills required to utilize the resources of their habitat, concluding that "this fishery is very difficult to all but these Indians who are trained to it from childhood."[7]

Following the departure of the Jesuits in 1696, information about the activities at the Sault is limited to a few comments by occasional visitors and traders. The advent of American and British traders in the late 18th century and the later increase in travellers and visitors led to more frequent descriptions of the fishery and the techniques of fishing and fish processing. Plentiful and reliable as the seasonal runs of fish were in the Lake Superior region, allowance had to be made for the provision of food throughout the lean winter months. Alexander Henry, who travelled in the region in the 1760s and 1770s, reported that the Indians smoke-dried large quantities.[8] Peter Grant provides even more detail, observing that the Saulteur preserved their supplies "by opening and cleaning the fish and then carefully drying it in the smoke or sun, after which it is tied up very tight in large parcels wrapped up in bark and kept for use."[9] John Johnston considered the fish at the Sault to be "the richest and best flavoured ever found in fresh water" and "are the chief support of both the Indians and white people here."[10] John Askin noted that "the Indians live entirely on fish. They even make their mokasins with the skins of sturgeon & Lace their Snow Shoes with the same skin & Skin the Muskelonge for the same purpose."[11] These traditional practices continued well into the 19th century and in 1885 the U.S. commissioner of Fish and Fisheries reported that "As late as 1865 crude smoking and drying frames, covered with cedar strips and hung with whitefish, were not an uncommon sight along the bank of the river in the vicinity of the rapids."[12] By 1885, however, the native fishery was in rapid decline, the advent of steamships and railroads facilitating the trade in fresh fish for the urban markets of the Midwest. Where preservation was necessary, salting and icing had come to be the acceptable processes. Although the Indians continued the traditional mode of preservation for their own domestic consumption, smoking fish for the commercial market was no longer practised.[13]

Despite the importance of these prosaic activities, it was the picturesque and hazardous fishing activity that elicited most comment. There were two basic methods of exploiting the fishery at the Sault rapids. So plentiful were the fish during the spawning runs that some could fish from the safety of the shore. J. Carver visited the Sault in the 1760s and reported that

> nature has formed a most commodious station for catching the fish which are found there in immense quantities. Persons standing on the rocks that lie adjacent to it may take with dipping nets, about the months of September and October, the whitefish...at that season together with several other species, they crowd up to this spot in such amazing shoals, that enough may be taken to supply, when properly cured, thousands of inhabitants throughout the year.[14]

2 Detail of W. Armstrong's Indians fishing on St. Mary's River. (Public Archives Canada, Picture Collection)

The more venturesome and more skilful fishermen, especially the local Saulteur, developed techniques that allowed a further exploitation of the available resource: a specialized two-person canoe, push-pole ("pique-de-fond") or paddle for steering, and poled dip net.

Anna Jameson was impressed by the intrepid and skillful performance of the fishermen, commenting:

> I lingered for a while on the burial-ground, looking over the rapids, and watching with a mixture of admiration and terror several little canoes which were fishing in the midst of the boiling surge, dancing popping about like corks. The canoe used for fishing is very small and light; one man (or woman more commonly) sits in the stern, and steers with a paddle; the fisher places himself upright on the prow, balancing a long pole with both hands, at the end of which is a scoop net. This he every minute dips into the water, bringing up at each dip a fish, and sometimes two....I never saw anything like these Indians. The manner in which they keep their position upon a footing of a few inches, is to me as incomprehensible as the beauty of their forms and attitudes, swayed by every movement and turn of their dancing barks is admirable.[15]

Visiting artists such as Schoolcraft, Catlin and Armstrong depict the swarming of the canoes and fishermen in the rapids and complement and verify the literary descriptions of earlier travellers.

Even as visitors recorded and romanticized these activities, the system was undergoing change. As early as the 1830s, some had seen

3 Fishing with specialized two-person canoes, push-poles or paddles and poled dip nets. (Parks Canada, Office Collection)

commercial possibilities in the fishery, which meant the eventual decline of the native use of the important resource. George Catlin reported:

> It has been found by money-making men to be too valuable a spot for the exclusive occupancy of the savages...and has at last been filled up with adventurers who have dipped their nets till the poor Indian is styled an intruder; and his timid bark is seen dodging about in the coves for a scanty subsistence, whilst he scans and envies insatiable white man filling his barrels and boats, sending them to market to be converted into money.[16]

On both the Canadian and American sides of the St. Mary's River commercial fishing was pursued vigorously during the 19th century. W.C. Bryant described the introduction of the new technology of seining associated with the new commercial organization of the fishery.[17] Despite these trends, the traditional mode of fishing the rapids continued and as late as the 1880s the U.S. commissioner for Fish and Fisheries could confirm the continued vitality of the native fishery. Having itemized the seine, gill net, pound net and ice fisheries, his report referred to 12 canoes on the American side and 6 additional ones on the Canadian shore, the 12 American crews alone landing some 75 000 pounds of whitefish in 1885.[18]

Other developments at Sault Ste. Marie in the 1890s disrupted the ecological base. The rapids formed the ecological niche of the Sault fishery and they were being transformed by two developments. The growing urban and municipal demands for hydroelectric power redirected and diminished the flow of the waters while, even more dramatically, the construction of the Sault Ste. Marie canal in 1893 completely altered the fluvial geomorphology of the locality and the basis of the fishery.

The beginnings of resource development in the mid-19th century, particularly the development of minerals, led to negotiations with the various Lake Superior tribes. By the Robinson Treaty of 1850 the Batchawana band reserved various lands including a fishing station at Whitefish Island; most of these reserved lands were surrendered for sale by Treaty 91A in 1859 but Whitefish Island was retained. Even after the canal was constructed, the pressure on the remaining lands continued. The access to the water power of the rapids attracted the interest of the Lake Superior Power Company in 1899. Hugh Hamilton, the solicitor for the company, petitioned Minister of the Interior Clifford Sifton for the surrender and sale of Whitefish Island by the Batchawana band for "hydraulic power development and erection of industrial works."[19] The next year Hamilton, now in his capacity as solicitor for the Ontario Hudson's Bay & Western Railway Co., applied for land on Whitefish Island for the development of a rail terminal.[20] Both of these enterprises were part of the growing industrial empire of Francis H. Clergue and he fully appreciated the potential of Whitefish Island. Asked to report on the nature of the Indians' occupation of the island, the Indian agent William van Abbott identified a seasonal population of only "ten habitations" but expressed concern that there was no suitable place for their relocation.[21] Writing from "Whitefish Island, Sault Marie, Ont., c/o J.C. Boyd, Supt. Ship Canal," Peter Kahgayosh provided an Indian perspective of the process in his plea that "I trust that this sale will not be made as it would cause great distress to us poor Indians and request that you will give me assurance that no sale will be made to stop my worry."[22]

Peter Kahgayosh was not allowed to stand in the way of Clergue and modernization. Armed with the appropriate clauses of the Railway Act and the Indian Act, all of Whitefish Island was appropriated for the national good and better operation of the Algoma Central Railway Co., the Pacific & Atlantic Railway Co. and the Ontario Hudson's Bay & Western Railway Co. Although these grandiose plans failed to materialize, they did displace the last of the Batchawana band from the rapids at the Sault.

The Fur Traders

The Montrealers

The first European contact with the Sault was that of the French explorers following the natural waterways west. After the recognition of the importance of the locality to the regional economy and social organization of the Indians of the area, it was natural that French missions soon followed. A permanent presence had been established by 1668 and continued until 1696. Moreover, in 1671 an elaborate ceremony marked the formal statement of the claim to the region advanced by one Simeon-Francois Lusson, Sieur de Saint Lusson, on behalf of the French Crown.[23] The rhetoric of the pageant referred to the glories of the Christian God and Christian king, but the reality of the claim was the introduction of a political and economic presence that was to have profound impact on the locality.

French fur traders were quick to avail themselves of the facilities established by the mission and the labour and hunting skills of the native population of the area. But the fur trade had an impact on the local cultures long before the actual arrival of the traders. Both furs and fish moved to the posts established east of the Sault. After establishing a local base, the Saulteur acquired the roles of middlemen between the French and the Dakotas of the western plains and of fishermen providing rations for the hunting operation. The 1696-1713 period witnessed a weakening of the northern operation. Fur-trading licenses were cancelled and a new policy introduced of attracting Indian sellers to travel directly to Montreal. The economic failure of this mode of operation, together with the political resolution of Anglo-French claims by the Treaty of Utrecht in 1713, stimulated a new western initiative. Between 1717 and 1731 Vaudreuil established a string of "postes du nord" from Lake Superior to Lake Winnipeg and the commercial significance of the Sault and its people was reaffirmed.[24]

Despite this activity, there are few records relating to the Sault during this period. In 1734, however, the Sieur de la Ronde, a Knight of the Order of St. Louis, arrived at the Sault to prospect for minerals. A sloop was constructed at St. Louis Harbour, now known as Pointe aux Pins.[25] After seignioral rights were granted at the Sault to De Bonne and Repentigny in 1751, a potentially new and more permanent French activity was introduced. It, however, was terminated by the outbreak of the Seven Years War in 1756 and the departure of the Seigneur Repentigny in that year.[26] Despite the war, at least some trading continued and as late as 1758 it was reported of Sault Ste. Marie that "the Saulteux do their trading there — there came from it about a hundred bales [of fur] annually."[27]

Although the French surrendered to the British at Quebec in 1760, it was not until 1762 that the British appeared at the Sault to occupy the French post on the south shore of the St. Mary's River. This first British presence was short-lived for the troops were withdrawn to Michilimackinac after the post at the Sault burned down. But the British still controlled the area and in 1765 Alexander Henry was granted the monopoly of the Lake Superior fur trade.[28] Henry's control of this extensive area was soon challenged by the numerous independent traders. Such was the disruption caused by the unregulated competition that a group of them agreed to cooperate in the exploitation of the region's resources, an agreement that was renewed for several years, culminating in the formal agreement to form the North West Company. Signed at Grand Portage in 1784, this document reaffirmed the dominance of the Montreal interests in the area, and the importance of the portage around the Sault in their system of communications.[29]

In October 1784 the new company petitioned the governor-in-chief of the Province of Quebec, Sir Frederick Haldimand, that they be allowed "to build a small vessel at Detroit to be sent to St. Marys with a part of the provisions for the purpose of getting her up the Falls, to transport the same over Lake Superior and to remain upon that Lake to be employed every summer in the same service."[30] Although the company did construct the 45-ton sloop *Beaver* for this purpose, it failed to haul it around the rapids to the upper lake. Not to be defeated, an even larger vessel, the 75-ton sloop *Otter*, was constructed in situ above the rapids on the shores of Lake Superior. In this way the original objective was attained; two independent systems of water transport were established on Lake Superior and Lake Huron linked by the indispensable, extended and difficult corridor at the Sault.

Although this system was functionally adequate for the needs of the fur trade, there were some who recognized the potential for a more efficient means of bypassing the rapids. Thus, Gother Mann of the Royal Engineers, while engaged in a survey of possible military sites to replace those that would be lost when Britain withdrew from the south shore, also commented on the operation of the existing transportation system at the Sault: "Canoes pass the rapid by *going up* quite light, and by taking out a part only of their loading to *come down*. There is a portage on the South Shore of about half a League in length..."[31] Moreover, as an engineer he also recognized the possibilities for improvements. The loss of the portage on the southern shore could be accommodated by the construction on the northern shore of "a road for Carts; which would also be shorter than the present Portage on the other side." In addition he proposed a more imaginative development for the north side where "one of the Channels passing between the Islands and the Main Shore might perhaps be capable with the assistance of Locks, of being converted into a navigable Canal." The relocation of the portage arrangements, be they by road or by canal, would benefit also from the potential for developing a harbour in the sheltered water at the "Foot of the Falls...where the necessary Wharfs or Quays might be erected and is therefore in this respect rather more convenient for Vessels to come to than on the other side." Jay's Treaty in 1794 allowed the North West Company 2 years to withdraw from the south shore and establish itself in British territory to the north.

The new trading post and surrounding lands were surveyed in 1797 by Theodore de Poncier and a year later the North West Company had partially implemented the Mann suggestion by constructing a small canal system which could accommodate "boats and canoes." In 1798 the North West Company had attempted to regularize their claim to the newly occupied lands and on August 10, 1798, they treated with the "chiefs & old men of the Chipeway at the Falls of St. Mary."[32] Having received the "consent & advice of the men of our tribe" and in return for two pounds sterling and other goods and considerations, the chiefs granted to the North West Company a 10-mile-by-10-mile extent of territory confronting onto the "falls of St. Mary's" and running inland from Grass Point below the falls and Pointe aux Pins above the falls.

Such an initiative by the North West Company threatened the interests of other traders, however, and in 1799 Phyneus Inglis petitioned the Duke of Portland that "for a roll of tobacco and some spirits for their [native's] rights" the North West Company intended to "exclude all others from forming any establishment" at the Sault.[33] The petition argued that all traders "must have storehouses on this Strait for the protection of their property" and that government "should reserve four or five leagues on the Strait and entrance of each Lake, leaving it free for all who engage in the Indian trade to make such Establishments there as they may find necessary." The Duke agreed, forwarding Inglis's letter to Lieutenant-Governor Hunter, adding the opinion that "it must be very much for the benefit of the Fur Trade, that about four or five leagues or perhaps the whole of the land along the Strait in question should be for ever retained in the hands of the Crown."[34] But the dispute concerned more than the land portage. A petition from the X.Y. Company in 1802 complained that the North West Company had improved and converted a part of the river channel "into a Species of Canal or Dam on the lower end of which they have erected a Saw Mill" and along which they claimed exclusive rights of navigation.[35] The X.Y. Company was opposed to this as it "considers Water Communications in the nature of Public Highways, and as such will use these when they see fit" and they also argued for the right to construct storehouses and wharves at the upper and lower ends of the portage. The counter claim by the North West Company provides further details of their improvements at the rapids, itemizing a 55-foot-wide portage road, a 3000-ft-long canal, a lock with a lift of 9 feet, and "a saw mill, store houses and other necessary buildings for facilitating the navigation of the said canal." They claimed that the canal was private property and that use by others would require the imposition of "an adequate toll." In 1802 Captain R. Bruyères, R.E., was despatched to survey the lands and facilities at the Sault and he provided a report on the specifications and mode of operation of the first canal there:

Close to the shore a Lock is constructed for Boats and Canoes, being 38 feet long 8 feet 9 inches wide. The lower Gate lets down by a Windlass; the Upper has two folding Gates with a sluice. The water rises 9 feet in the Lock. A leading Trough of Timber framed and Planked 300 Feet in Length 8 feet 9 inches wide 6 feet High supported and levelled on Beams of cedar thro the swamp is constructed to conduct the Water from the Canal to the Lock. A [Road] raised and Planked 12 feet wide for cattle extends the whole length of the Trough. The canal begins at the Head of it which is a Channel cleared of Rocks and the projecting Points excavated to admit the passage of canoes and Boats. This Canal is about 2580 feet in length, with a raised bridge or Path Way of round Logs at the side of it 12 feet wide for oxen, to track the Boats.[36]

The provision of lots for the facilities required by other traders, the survey of a road by Lieutenant Brice, and the construction of yet another portage road by the X.Y. Company parallel to that already established by the North West Company established the Sault as a focal point on the east-west commerce of the Laurentian system.[37]

In 1804 the North West Company amalgamated with the X.Y. Company, ending much of the conflict over their respective rights to the portage at the Sault. Eight years later, the Sault witnessed the replacement of commercial competition with military hostilities when war broke out between the United States and Great Britain. Merchants from Sault Ste. Marie participated in the capture of the American fort at Michilimackinac and the Americans retaliated by seizing the Sault post in July 1814.[38] Whether at the hand of the company's agent or that of the invading American forces, the facilities at the North West Company's trading post, including the lock and mill, were destroyed. But within a year the company had reoccupied the site and reinstituted their trade with the northwestern territory until their amalgamation with the Hudson's Bay Company in 1821.

The Men from the Bay

After the amalgamation of the two former competitors, the continued importance of the Sault to western commerce came into question. The traffic via the Great Lakes—St. Lawrence system was rerouted by the shorter connection to Hudson's Bay at York Factory. Fort William lost its significance and the Sault and other posts along the Montreal route were similarly affected.

Indeed, initially, the commercial presence at the Sault was threatened by the recognition of the military significance of its location astride the new international "line" established by the Treaty of Ghent in 1815. The H.B.C. post there was considered a possible military base, "which they have offered to Government at a very moderate price."[39] By March 1824, Colonel Darling reported to the agent of the company, "I have received His Lordship's orders to acquaint you that he has no objection to authorize their being purchased for Government at the price of Two Thousand Pounds Currency... for the whole of the Buildings,

Wharves, etc. as therein described; the premises connected with the Establishment reverting to the Crown on the purchase of the buildings."[40] A ground plan of the post at that period depicts some 17 structures including a sawmill, a shed for building boats, carpenter's shop, houses, various barns and outhouses, company store, wharves and an oven for charcoal.[41] The military were also interested in "the several small pieces of Ordnance and some field Guns with their appurtenances...which they are about to dispose of," as they were concerned "they should pass into the hands of the Americans."[42] No extensive military presence was established at the Sault, however, and the British allowed the American Fort Brady uncontested control over the approaches to the portage.

For the next half century the trading post at the Sault functioned as an element in the H.B.C.'s system of trade and communications. For the first decade the "New Establishment," as the Hudson's Bay post came to be known, continued to supply provisions for interior posts, items such as potatoes and fish being obtained locally and other commodities from farther afield. Similarly, although Montreal was replaced by London as the organizational centre of the fur trade, some trade and communications still moved east and west via the portage at the Sault. Finally, although the Sault itself was never a centre of the fur trade, by mid-century a limited amount of fur buying, an increasing trade in the export of salted fish, and the sale of goods to a burgeoning local population came to be new functions for the Hudson's Bay outlet at Sault Ste. Marie.

On September 1, 1824, Angus Bethune, the new Hudson's Bay Company factor at Sault Ste. Marie, recorded in his journal that he had "received charge" of the post from Mr. McBean.[43] Located across the creek from the old North West Company post, the "New Establishment" had yet to be developed. By the spring of the following year, however, Bethune could report much progress:

> at this date, there is completed of the new Establishment a Store of 60 feet long and 24 wide, containing three floors. The first one is divided in the following manner. One half is appropriated for Lumber Liquors and other articles not subject to damage from vermin. The other half is divided into two shops and a dairy....The largest one contains all the goods brought from Fort William in the month of June last. The other one is intended for the use of the Post exclusively thereby preventing all communication with it by the servants, daily laborers, Fishermen.... The middle Floor, or first Story [sic], contains all the Provisions, with many other articles. There is a clear space of the whole extent of the building, that can contain a large quantity of goods. The access to it is by a Gallery on the outside of the building fitted with skids & Ropes, for Rolling the Goods up or down, when Boats Land & unload within a few yards of the Doors. The third floor or Second Story [sic], is also a clear space of the whole extent of the building, intended for dry Goods, Furs, or any other articles...[44]

Other buildings completed during this first season of occupation included a "dwelling house." Future developments "required to complete the establishment" included a kitchen, another house, carpenter and smith shops, stables and root house. The post had two stockaded "spots" for the cultivation of potatoes, but had lost its arable land at the old establishment. Noting this, Bethune reported that "rather than be at the Expense of Clearing Land on this West Side, I should prefer purchasing Oats and Barley" and recommended the sale of the livestock, retaining only two horses, a bull and two cows. He also reported on the general state of disrepair of the "Old Establishment," except for the mill which "can be kept in good order as much from the new place as the old one," and concluded that "The New Establishment will be compact and convenient and last many Years."

The shift from the "Old" to the "New" establishments meant more than a change of site and construction of new facilities. The amalgamation of the two companies required a new orientation of the trade patterns. Just as the "Northwesters" had used the watershed of the Great Lakes for their trade, the "Honourable Company" used that of the Hudson Bay drainage. Because of the dominance of the H.B.C., the importance of the portage at the Sault and the maintenance of a post could not be justified. Accordingly, Bethune's first report also recommends a reduction of the operation "to 2 labourers, 1 servant and 1 Clerk, with a Commissioned Gentleman if one is thought to be necessary." The annual report for the following year recommends that no goods be sent for the "Indian Trade" as pursuing the fur trade would "be adding Expence to this Establishment in place of reducing them."[45] The next year, 1828, the situation was worse, Bethune reporting, "The trade of furs at this place is not improving but getting worse."[46]

Bethune had previously anticipated at least some trade with the local population in the "Suburbs" and the American population across the river, having been advised that "it would be well if the Hudson's Bay Company imported some fine clothes and other Woolens suitable for their use, that they would have the whole Custom of the Garrison...."[47] But this was not to be. By 1829 the British had withdrawn their garrison at Drummond Island, the Americans following suit by reducing theirs at Fort Brady to 3 officers and 63 men, none of whom purchased from the company. As for the local custom, Bethune complained that "The Population is decreasing Rapidly from Deaths and Emigration to other quarters. I think that our side of the line has decreased more than 1/3rd since last summer." George Simpson was confident in Bethune's abilities despite these developments.[48]

Simpson pointed out trade was "merely a Provision Depot for the Southern Department and passing Brigades," a description that perhaps unintentionally diminished its importance. In some ways it was merely one more station along the line of communication from the H.B.C. and other western posts to Montreal. Montreal had been replaced by London as the headquarters of the fur trade, but communications still moved east-west via the portage at the Sault. The post journals referred frequently to the arrival or departure of the "Canoes from the North" or the "Montreal Canoes."[49] Another frequent element of this extensive communication system were the "small" canoes carrying dispatches from the Northern Department to London via Montreal and in winter, the arrival of "the express from Moose Factory." One such dispatch provides a good example of the system's expediency and the herculean efforts of the men on which it relied:[50]

Posts	Arrive	Depart
Fort Alexander	Dec. 23 (8 AM)	Dec. 23(10 AM)
Fort Frances	Jan. 3 (5 PM)	Jan. 4 (7 AM)
Fort William	Jan. 14 (2 PM)	Jan. 14 (4 PM)
Lake Nipigon	Jan. 21 (2 PM)	Jan. 21 (3 PM)
Long Lake	Jan. 25 (3 PM)	Jan. 25 (3:30 PM)
Pic	Jan. 30 (10 PM)	Jan. 31 (daylight)
Michipicoton	Feb. 6 (12 noon)	Feb. 7 (daylight)
Batchewanna Bay	Feb. 11 (evening)	Feb. 12 (daylight)
St. Mary's Depot	Feb. 13 (2 PM)	Mail left in evening

The post was a crucial link in the "Provision Line" for the posts in the Lake Superior and Lake Huron districts. The traditional operation of the "Honorable Company" relied upon the annual arrival of "The Ship" at the H.B.C. headquarters on the bay. Because of development of the western interior of the United States and the recognition of the possibility of supplies coming from that quarter, the Sault again became the lynchpin in a continental wide transport system. The portage again linked the two separate systems of water transport on Lake Superior and the lower lakes. The factors at the Sault kept up an active correspondence and commerce with American traders in Detroit, Cleveland and other American centres. As early as 1826, Bethune had recognized that provisioning the posts occasioned "much inconvenience to business of head office" and recommended that "two batteaux carry each sixty barrels should be made at this place, and ready to launch on the opening of the navigation next spring."[51] Giving the example of supplying Michipicoten, he argued the cost of freight for two trips per boat, per year for a total load of 240 barrels as 45 pounds sterling, or less than four shillings per barrel. His recommendation was approved. Bethune launched his new boats for the Lake Superior trade in May 1827. In June he reported "taking up the Boats in the Channel from the Old Fort to the upper wharf," and in July "The Lake Superior men were employed taking the provisions to the old wharf, from which place Raymond carted them across the portage" to be loaded into the Lake Superior boat accompanied by one of the new boats manned by "St. Mary's Inhabitants."[52] Henceforth the annual responsibilities of the post would include the unloading, loading and, in some cases, portaging around the rapids of the "goods and provisions" for the boats for Lake Superior, Lake Huron, Michipicoten, Nipigon, Drummond Island and other posts. Increasingly, the arrival of boats on the "American side" was more important to the operation of the system than the annual "Ship" at the Bay.

The "New Establishment" was an important link in the H.B.C. system of provisioning the interior posts. In particular, the native fishery was closely integrated into the supply of provisions for the posts along the line and increasingly became part of the commercial interests of the company.

The initial emphasis was on exploiting the fishery for provisions, but by October 1835 the emphasis had shifted to commercial development. The new factor, Nourie, reported that some 70 barrels were ready for the trade.[53] By November 955 barrels of whitefish and 6 barrels of lake trout were shipped to Mr. Oliver Newberry of Detroit. This was the year that Simpson had advised Nourie the company proposed establishing a fishery on Lake Superior "with the view of meeting the demands of the American Market in the article of salted fish."[54] This new venture required the construction of a 50-ton vessel for navigating Lake Superior and by September 16, 1837, Nourie could report back that the new vessel, *Whitefish,* had been constructed at Sault Ste. Marie, launched and was heading for Fort William "before the stormy season commenced."[55] At the commencement of navigation in 1838, Nourie advised his superiors that "the vessel (Whitefish) will be disposable for any fisheries that can be carried on with advantage."[56] By August Nourie reported that "the fisheries have been very successful," 276 barrels of fish at $8 per barrel having been sent to market, and that Mr. Gidings was interested in entering into business with the company.[57] Clearly, the "Honourable Company" was developing profitably.

If only marginally involved in the fur trade itself, and if only of regional importance in the continental communications system, the company did maintain a presence at the Canadian Sault. In particular, the company "shop" was both an outlet for goods and a buyer of furs from local trappers. Indeed, by 1835 the economic fortunes of the post showed some signs of improvement indicated by the opening of the "sale shop," the development of the "fish trade," and the "experiment last year of bartering for furs."[58] Factor Nourie reported his intention to "extend that and any other branch of business" that promised "a high profit for the goods." The provisioning of the interior posts continued to be important, Nourie reporting in 1837 that "the forwarding business of this Outfit has been completed in good season and I believe satisfactorily to the Gentlemen in charge of the contiguous Districts. There is now an ample stock of provisions in store for next Outfit."[59] Since 1832 the costs of this operation were charged to the operation of the Sault post, but Nourie had applied himself to the task of "turning this Outfit to profitable account."[60] Noting that the Lake Superior forwarding operation "swallowed up all of the profit made by the small sale shop," he had every expectation that "the Balance may soon be changed from a loss to a gain." A decade later, however, a new factor, W. McTavish, complained in his private correspondence that "our shop here is a miserable humbug there being little now selling in it & on what is sold not much profit is made."[61]

Despite the continued attempt to maintain a store and some buying of furs, by 1865 plans were being made to close the Sault post. Simpson recommended that the place "be sold before the buildings become useless which they will soon be if they are not put in good repair."[62] But the company did not have legal title to the 1200 acres which they claimed at their post at the Sault. Simpson recommended that if it proved to be "impossible to get the Government to come to a settlement by allowing the claim for the 1200 acres to ask to be treated as the other squatters were for the forest lots & for the rest of the lands" at a price of 70 cents per acre.[63] Because of a disputed claim to the land, the deterioration of

4 Hudson Bay Company post at Sault Ste. Marie by William Armstrong. (Metropolitan Toronto Public Library)

the physical plant and the economic viability of the operation being questioned, the late 1860s witnessed the final demise of the Hudson's Bay Company Sault operation. Although company personnel continued at the Sault, the establishment was closed and a vital phase in the history of the region ended.

The Close of the First Contact

Clearly, the rapids constituted a key feature in the landscape to the first societies of the Sault. To the "Bawatig" they were an important element in their resource base and developed a crucial dimension of their cultural distinctiveness. As proprietors and principal exploiters of the fishery there, they were to become participants in, and observers of, subsequent developments at the site of their traditional home.

For the fur traders, the importance of their successive establishments at the Sault was in facilitating movement around the barrier of the St. Mary's rapids, especially when such movement was focussed on the terminus at Montreal. The union of the North West and Hudson's Bay fur-trading companies in 1821 dealt the first major blow to this "raison d'être." Its pivotal position on the line of communications between the Canadas and the posts to the west and north of the Great Lakes became redundant. At least initially the reorganized Hudson's Bay Company focussed its strategy around the old north-south system with its terminus at Churchill on the bay and its major distributive centre at Fort Garry. Eventually, however, the economics of transport and the logistics of supply dictated that the Sault emerge as the break-in-bulk point for the distribution of goods from the south and east to the posts in the Lake Superior and Lake Huron districts to the north and west. To some extent, therefore, the old east-west function continued.

The former British North American and British sources of supplies and commodities for the posts were replaced increasingly by United States producers. The rapidly expanding American railway network rendered the old dependence on transatlantic suppliers of goods redundant and the Sault was reduced to a minor trading outpost of the company. But the expansion west of the new Canadian state, the growth of commerce, and the emergence of the new resource demands of an industrializing world focussed renewed attention on the Sault and its rapids.

II BYPASSING THE RAPIDS: THE CASE FOR THE CANALS AT THE SAULT

The "Soo" at Mid-Century

On December 4, 1845, Commissioner of Crown Lands D.B. Papineau directed "Deputy Provincial Surveyor Alexander Vidal to Survey a Town Plot and Park lots at Sault Ste. Marie."[1] In June 1846 he was charged further with the duty of producing a plan of the subdivision of the town plot, marking on it "the ground which should be reserved for the sites of churches, schools, markets, squares, burying grounds & other public purposes; and the buildings & clearings of the settlers with their names."[2] His instructions concluded with the directive to "Consult the wishes of the inhabitants in naming the street & squares." By the end of the summer of 1846, Vidal had laid out a town plot of 324 acres, 3130 acres of park lots, and some 15 streets. After the incorporation of the town in that year, Sault Ste. Marie became prominent among the communities along the north shore.

Despite its post office and branch of the British North American bank — both managed by the Hudson's Bay Company agent — the community still comprised only 500 persons focussed on three elements: the company post with its skeletal staff and employees, the French-speaking Métis population clustered around the mission in the "suburbs," and the Indian village on Whitefish Island. The 1861 census recorded 898 persons of whom 653 were native Indian, 112 were British and 75 were French Canadian.[3] Such was the general state of the economy at the Sault that this population was to decrease to 879 in 1871 and to 780 in 1881.[4]

Apart from the descriptions of those seduced by the romantic image of the northern wilderness or by the promise of an as yet unrealized potential for resource development, the accounts of travellers to and residents of the Canadian Sault present a dismal view of the community in the mid-19th century. Thus, William H.G. Kingston, an Englishman who wrote travel books, visited the Sault in the early 1850s and commented on the Honourable Company's facilities at the time of his visit:

> The post is a great white washed edifice, partly built of logs and partly frame, and surrounded by a paling. More in the interior it would be a stockade. The log building is infinitely the most comfortable, and when planked over looks very neat and most effectively keeps out both cold and heat. All the dwellings at the Hudson Bay Company's post are of log; but their stores are frame buildings.[5]

Kingston also wrote about his impressions of the larger community, the major resource at the time being the nascent tourist industry, and he commented on the "several hotels, which are resorted to in summer by visitors both from the States and Canada, who go there to enjoy the cooling invigorating breezes from Lake Superior."[6] The Hudson's Bay Company viewed these developments with some disfavour:

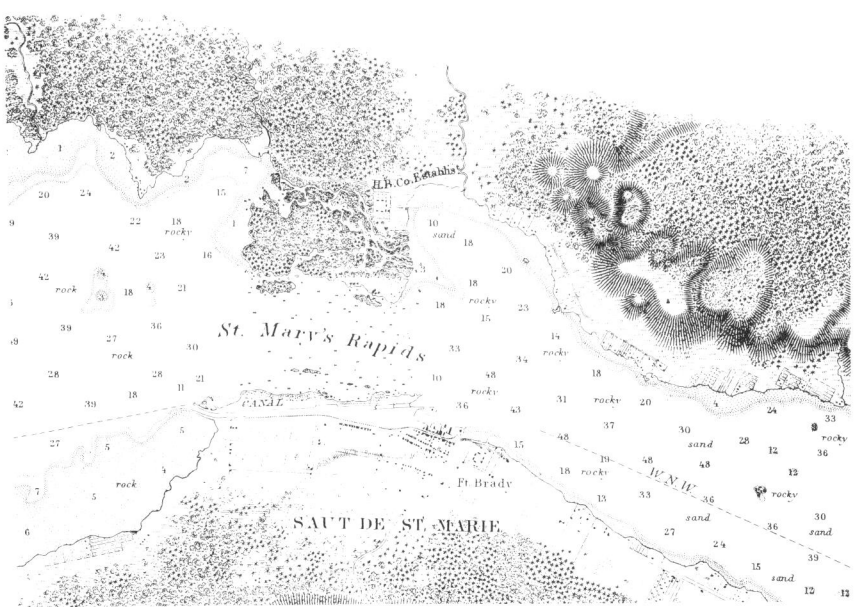

5 River St. Marie from Point Iroquois to East Neebish. (Bureau of Topographic Engineers of U.S. War Department)

> Dr. Rae is going on another journey to the Arctic Regions, it may be politic on the point of the Company to place before the rest of the world their exertions in adding to the known geography of our planet but I fear their endeavours to this respect will tend to curtail our dividends and these hyporborean regions are now becoming so well known to the civilized world that, old as we are, I should not be surprised if we lived to see these regions becoming as fashionably a summer resort to the travelling community as Saratoga and Cheltenham now are.[7]

But at mid-19th century the Sault was far from attaining the standing of a major spa. To be sure, there were other facilities, if somewhat consolidated in the personnel and establishment of the H.B.C.'s post, and Kingston referred to the "the planked residence of the police-magistrate, custom-house officer, post-master — one individual, a very courteous and intelligent gentleman, representing all the official dignity in the place."[8] Kingston was not, however, as impressed by the diligence and industry of the other residents of the place, referring to them as "half-casts, the offsprings of Indians and French Canadian voyageurs and are a degenerate, dissolute, ill-conditioned set. They work when driven by necessity to find food." Across the river, the American Sault provided him with an appropriate contrast of industry and vitality with "a large steamer...near the quay [below the rapids]; and as we looked up the watery hill, we could see another above the rapids letting off her steam having just come in from the many regions of Lake Superior."

6 John Prince, ca. 1866. (Notman Photographic Archives, McCord Museum, McGill University, Quebec)

John Prince, the elected member of the legislative council for the western division, 1856-60, and judge of the Algoma district, 1860-70, visited the area in 1853. He, too, was singularly unimpressed, commenting, "I think the 'Soo' a wild & horrid & inhospitable place. Should not like to live there."[9] But live there he did and after 7 years residence as the most powerful official in Algoma he retained his distaste for "this accursed 'Soo'..."[10] Another resident, the Hudson's Bay Company's agent or "Bourgeois," William McTavish, also viewed his appointment there with some distaste, complaining that he had been sent to the Sault as a punishment for his disobedience. He commented:

> The Sault on the American side — for on the British it consists of a few miserable hovels inhabited by half breeds who vegetate on white fish — is a village of considerable size. The inhabitants are a set of speculating knaves who have been drawn there some years by the expectations of making their fortunes from the

pickings of the copper mining business in Lake Superior of which directly or indirectly is the means of most of their livelihood....[11]

Apart from the contemporary prejudices and personal biases of these commentators, what were the reasons for the apparent disparity in development of the communities on opposite sides of the rapids? The answer lies apparently in differences in the political and economic developments taking place in the United States and Canada during this period.

Early Developments

The American Sault

South of the border, developments were more propitious for the American Sault. Following the American Revolution, this relatively isolated wilderness area became part of the United States, but the British declined to withdraw from substantial areas of the Old North West for several years. Not until the Jay Treaty of 1794 did the British agree to abandon these lands on the south side of the lakes, and not until 1805 was it organized as the separate territory of Michigan. As late as 1820 the non-native population of the entire territory was less than 9000 people. The 1820s were crucial to Michigan's history. In 1820 General Case negotiated a treaty with the Chippewa in the upper portions of the territory. After this was settled and after the construction of the Erie Canal in 1825 facilitating the western movement of population, settlers quickly poured into the territory. On January 26, 1837, the State of Michigan came into being, including within its bounds the rich resources of the upper peninsula to the south of Lake Superior, an area that became tributary to the American Sault.

The upper peninsula was rich in copper and iron and early state administrations, especially that of Governor Stevens Mason, were anxious to exploit this wealth. Copper and iron could be removed only by water transport along the south shore of Lake Superior. The most expeditious way to get the mineral wealth to the industrialized regions in the East was through the Great Lakes—Erie canal system. For such a system to work, a canal was needed at Sault Ste. Marie, it being immaterial whether the canal was in Michigan or Western Canada.

In 1839 a three lock system was started at the American Sault, but the state's contractors were forced to halt work after a dispute with the military at Fort Brady. Why the military objected to a canal is unknown, but that opposition by no means ended the issue; pressure for a lock at the Sault became more intense during the 1840s as Michigan hosted a copper rush and presided over the opening of several iron mines. In the interim, the traditional portages around the rapids served the needs of the ever-increasing trade. Moreover, between 1839 and 1845 the schooner *Algonquin*, the propellor steamer *Independence*, and the side-wheeler *Julia* were all manhandled across the portage to serve as carriers on Lake Superior.[12] Such heroic gestures may have demonstrated the energy and determination of those involved, but they could not possibly be counted on to solve the problem of transporting the considerable resources of the upper peninsula. Unfortunately, the early Michigan state administration had been excessively expansionary since 1837 and by the 1840s the state coffers could not support the expenditures required for canal construction at the Sault. Accordingly, for most of the 1840s, Michigan's canal strategy was focussed on Washington, but without the political clout to impress its economic policies on a southern dominated Congress and Senate. The project had to await more propitious times.

If the concept of a canal on the American side did not engender U.S. support outside Michigan, the Canadians were very aware of the benefits to the American economy if one were built on the Canadian side. The 1852 Keefer report emphasized that if a Canadian canal were constructed at the Sault, "Three-fourths of the business to be done upon it will be, as on the Welland, between American and American Ports" and would serve the needs of Michigan's 2500 miners, 15 companies in production, and another 22 mining operations undergoing development.[13] The *Niagara Chronicle* warned the government that it might well lose a first-rate opportunity: "The Americans are not asleep about a ship canal round the Sault, and our Government should look out or they will be superseded in the great object which has been so repeatedly urged upon them."[14] To prove its point, the *Chronicle* quoted a Cleveland newspaper which strongly supported an American canal at the Sault:

> When completed this canal will enable steamers of the largest class to run from Buffalo to the headwaters of the lake without any transhipment — an improvement that will quadruple the trade of this immense lake in less than two years...In the way of mineral wealth it [i.e. the Lake Superior shore] acknowledges no equal anywhere; heretofore all this wealth has been landlocked. Owing to the Falls of St. Mary, all the products of Lake Superior have been so hindered and taxed, in order to reach the seaport, that it was almost impossible to make them pay expenses.[15]

The editor of the *Niagara Chronicle* was correct. The Americans were not asleep and had not been asleep. Throughout the 1840s Michigan's political leaders continued to lobby Congress for support for a canal. During this period, the economic logic for such a venture became stronger and by 1849, 96% of U.S. copper was mined in the upper peninsula, which also held vast reserves of high-quality iron deposits. Optimal exploitation of these resources was effectively blocked by the rapids at the Sault.

On August 25, 1852, President Millard Fillmore signed a bill to grant 750 000 acres of land in return for a canal at the American Sault. The operation was administered by the State of Michigan. The state considered a variety of bids and then let the contract on April 5, 1853. In conformity with the somewhat murky business practices that prevailed in U.S. transport construction at mid-century, the contract was assigned to the St. Mary's Ship Canal Company, a firm incorporated in New York. The St. Mary's Ship Canal Company was the construction arm of the group of men who obtained the original contract on April 5. The chief person involved in the construction was Charles T. Harvey and he was given only 2 years to complete the project. Harvey, a former salesman from Vermont, had been active in the canal's promotion and was to become the firm's general agent.

Work proceeded rapidly and efficiently. The project was costly for initially the entrepreneurs who built the canal estimated that it would cost between $260 000 and $403 000, and the final cost was $200 less than a million dollars. As many as 1600 men worked on the project, which began in June 1853. The work was finished on schedule and turned over to Michigan in May 1855. Harvey and his labourers presented the state with two locks in tandem that raised or lowered ships 18 feet. The canal was 350 feet long and 70 feet wide. The original locks were used until 1887. After Michigan assumed control in 1855, it appointed a superintendent who collected tolls that were used to maintain the facility. In 1881 the canal was transferred to federal jurisdiction and tolls were abolished. The American canal was a tremendous success, lockages increasing from 14 500 tons in the first year of operation to 284 350 tons in 1865 and 1 505 780 in 1875.[16]

The canal on the U.S. side had been built and was a success. The wealth of Michigan's upper peninsula could be exploited, but so could the resources on Canada's Lake Superior shore because the canal was opened to the commerce of British North America as well as that of the United States. Canada had let an opportunity slip by and the people at the Canadian Sault were bitterly resentful. William Kingston described their sentiments:

> The neglect of the Canadian Government in forming the canal is a sore subject with all the British acquainted with this region. The ground was actually surveyed, and found more practicable

7 American canal at Sault Ste. Marie by William Armstrong.
   (Metropolitan Toronto Public Library)

than that on the American side. Why it was not done seems a mystery. The Lower Canadian party are accused of throwing obstacles in the way of the work through jealousy of the advantages it would bring to the Upper Provinces, but that I should scarcely think possible. It is far more probable that the neglect arose from an ignorance of or indifference to the important advantages the undertaking would have secured to the country at large.[17]

The American Sault bustled and grew while its Canadian neighbour slumbered fitfully through the 1860s and 1870s.

The Canadian Sault

Despite the relative decline of the area and perhaps prompted by the American developments, in 1846 the government of the Province of Canada despatched Hamilton Hartly Killaly, a senior government civil engineer, to report on the feasibility of a ship canal on the Canadian side of the rapids at the Sault. Following his visit to the site, Killaly proposed a canal, a proposal best demonstrated by the maps and plans that accompanied his report. In 1846 the Province of Canada was administered by the government of William Henry Draper, a weak regime preoccupied by constitutional, factional, educational and bicultural

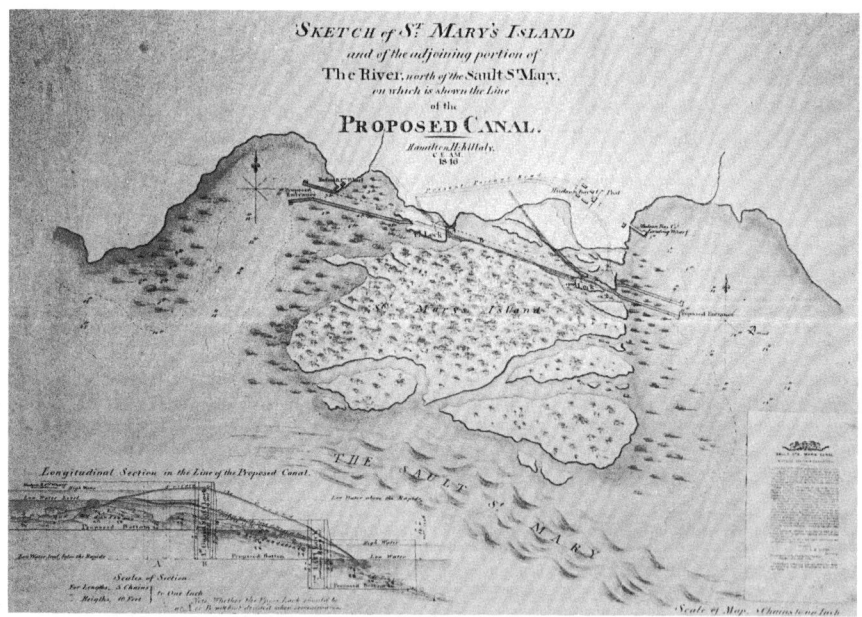

8 Sketch of St. Mary's Island and proposed canal, 1846. (Parks Canada, Sault Ste. Marie Canal Collection [SSMCC])

crises. It was not interested in broad questions of economic policy, nor had it any orientation towards the frontier or resource development. Killaly was probably sent to the distant Sault from Montreal because he was not acceptable as the Minister of Public Works. The 1846 probe cannot be taken as a serious initiative.

Much more serious interest was shown a few years later. Both Samuel Keefer, chief engineer in the Department of Public Works, and a parliamentary committee investigated the possibility of a canal at the Sault in 1852. The House committee reported after Keefer, but its report includes a precise statement of the economic strategy that lay behind the pressure for a canal:

> By cutting a canal round Sault Ste. Marie at once, we should secure the carrying trade of the Americans through our own canals to the Atlantic; offer to those who might be inclined to embark on either of the above named pursuits [mining and fishing on Canadian shore of Lake Superior], increased facilities, and consequently increased incentives; and probably accomplish the object of our present inquiries.... Every ton of goods or copper has to be transhipped and conveyed over a rude railroad, about a mile in length, at a very considerable expense to the public....[18]

In his report Keefer referred to his detailed notes concerning the Sault region and to appropriate charts of the waterway connecting Lake Superior with Lake Huron. You "cannot fail to observe," he explained,

> how simple and easy it is to remove all obstructions and open up a communication with Lake Superior. From *Gross Cap,* where the River may be said to take its departure from Lake Superior, to the head of St. Joseph's Island, where the North, or Canadian channel, leading off into the Georgian Bay, leaves the one leading to Mud Lake, which is connected with Lake Huron, the distance is 40 miles. The main navigable channel in its general character is deep and capacious, and presents but two places where the navigation is either stopped or seriously impeded by natural obstacles. The first is the Sault Ste. Marie, and the second, the bar in Lake George.[19]

The latter obstacle is dismissed as being easy to overcome with a simple "cut through the bar...." The other obstruction would require a ship canal, and it would have to be a large one. Keefer noted that some argued the future traffic of Lake Superior would be dominated by "propellors" and vessels of "cheaper build" rather than by large steamers. But he disagreed, arguing that a canal should be built to accommodate the largest vessels which, "if a way were once opened into Lake Superior for the large steamers, they would soon enliven its surface in considerable numbers." Keefer also argued that "Locks in my judgement should have a length of 350 feet in the chamber between the gates, 66 feet in width, and 10 feet in depth upon the sills." The fact that such locks would be larger than others in the Canadian route to tidewater was no argument against their construction: "Before 15 years shall have elapsed another set of Locks will be required on the Welland to meet the requirements of a trade which increases regularly at the rate of twenty per cent per annum."

The location and plan of the ship canal was straightforward: "The

Canal has been laid out upon a straight line, the shortest that can be drawn between the navigable portions of the Bays, above and below the Islands, thus passing nearly through the middle of the large Island on the Canadian side. Its length through the Island is 50 chains, but from end to end of Piers it is 95 chains." Keefer advocated "two Locks to overcome a fall varying from 17 to 19 feet," the finished canal to be "140 feet wide at surface, and 130 feet at bottom, wide enough for two vessels to pass each other in any part of it." The cost estimate for "a Canal with Locks of this size if ₤120,000." Keefer than gave cost estimates for two options: "a Canal, 120 feet wide, with Locks, 250 Feet x 55 Feet x 9 Feet is ₤100,000," whereas "a Canal, 140 feet wide, with Locks, 350 Feet x 66 Feet x 10 Feet is ₤120,000." He felt the cost was not a serious consideration because tolls would pay the interest charges on the capital investment.

In 1852 Francis Hincks was the dominant influence in Canadian politics. His administration was obsessed with transportation policy, but the focus was on railroads not canals. In the mid-1850s Canada's public credit was stretched to the breaking point in financing the Grand Trunk Railroad. A ship canal at the Sault was not nearly important enough within Canadian transportation policy to obtain government approval and funding. But it did have some support, especially in the West. The Goderich *Huron Signal*[20] and the *Owen Sound Comet*[21] argued strongly for the construction of the canal. Despite their enthusiasm, however, pressure for the project waned and the Canadian Sault lamented the government's failure to be imaginative and innovative in the face of the aggressive verve and vigour of their neighbours to the south.

Economic Initiatives

Despite the failure of the Killaly and Keefer reports to persuade government to invest in improvements at the Sault, events in the latter half of the 19th century underscored their arguments. The 1850s witnessed two developments that had substantial implications for transportation policy in the far west of the Province of Canada. First was the increasing interest and activity in the resource potential of the north shore of Lake Superior. Second was the major thrust towards the Prairies which, of course, changed the context of transportation policy from mere regional considerations to the transcontinental. Both developments emanated from Toronto and were closely tied to that city's rise to metropolitan pre-eminence within what was to become the Province of Ontario.

Intimately involved with both developments was Allan Macdonell (1808-88), a Toronto Liberal and entrepreneur. In 1846 he "obtained from the Government a license for exploring the shore of Lake Superior for mines...."[22] With other Toronto entrepreneurs he began to explore the area in 1847 and spent most of the next 10 years on the Lake Superior shore. The ultimate result was the formation of the Lake Superior Mining Company which extracted ores "successfully for several years." Macdonell had connections with other mining concerns as well. He was active in the Victoria Mining Company and became its president in 1856.[23] In 1865 he was appointed managing director of the Upper Canada Mining Company.

Like their counterparts to the south, these Canadian businessmen (Macdonell, his associates and others) were concerned with the transportation problems presented by the St. Mary's River rapids. In 1849 the Quebec Mining Company explored the possibility of a portage railway around the rapids.[24] Macdonell, his brother Angus and various associates were also interested in sorting out the transportation problem. In 1851 they applied for a charter to build a railway to the Pacific and also attempted to get a charter to build a canal at the Sault. Both bids were rejected. The attempt to get a charter to build a transcontinental railroad was probably simply a part of an anti-Hudson's Bay Company propaganda war; Macdonell and his associates did not have nearly enough capital to pursue these ventures.[25]

The opening of the canal on the American side in 1855 ended the transportation blockade, but subsequent development on the Canadian side remained minimal. Silver Islet, close to Thunder Bay, was discovered in 1868. It produced large quantities of silver in subsequent decades and stimulated considerable prospecting activity. The Bruce Mines were productive during the 1870s and various other mining operations produced limited amounts of wealth during the 1870s and 1880s. Between 1871 and 1881 the Algoma district silver production increased from 69 197 ounces to 87 000 ounces and in each year that constituted the total production for Ontario. In the same period, copper production apparently decreased from 1934 tons to 150 tons but again accounted for almost the total provincial production. This decrease in copper traffic was more than compensated for by the addition of some 1600 tons of iron to the Algoma district mineral production by 1881.[26]

Lumbering was another resource industry that was expected to become important locally. The Department of Crown Lands surveyed properties around Sault Ste. Marie and eastward as early as the late 1840s, continuing through the 1870s. Timber limits on the north shore of Lake Superior were sold from 1868 but activity was slight until after the 1870s. Near the Sault there was more lumbering activity. Ottawa valley lumbermen arrived at Grand Marais in the early 1870s and their interest was the square-timber trade. Sawmilling was well established on Sugar Island in the 1880s. These activities were important but resource development near the Sault or west of it was not of major importance until the last decade of the 19th century.

The second important development in the economic sphere was the emergence of a Toronto-based passion for westward expansion into the Prairies and to the Pacific. Again Allan Macdonell and associates, supported by George Brown and his influential Toronto *Globe,* were central to this movement.

Brown and the *Globe* espoused the cause of westward expansion as early as 1847[27] and in 1848 the *Globe* attacked the alleged rights of the Hudson's Bay Company,[28] whose territories, it was argued, were "capable of supporting a numerous population. This wide region nominally belongs to the Hudson's Bay Company, but in point of fact it does not seem to be theirs." In 1850 the *Globe* campaign ended for a while and the issue was virtually dead for several years. Late in 1856 the recently organized Toronto Board of Trade came to the aid of the movement to annex the Prairies to Canada. Allan Macdonell addressed a board meeting which duly resolved to ask the Canadian legislature "to ascertain what are the legal rights of the Hudson's Bay Company to the territory and exclusive trade claimed by that Company in the northern part of this Continent, and to pray them to adopt such measures as may be necessary to protect the rights of this province."[29]

Pressure for westward expansion continued. Macdonell and his associates worked hard to obtain a charter to build a line of communications to the Pacific. Finally, in 1858 they succeeded in obtaining a charter for the short-lived North-West Transportation, Navigation and Railway Company.[30]

Within the Province of Canada, and especially within Ontario, there was growing interest in the exploitation of the resource wealth of Canada's Lake Superior shore combined with a desire to expand into the Hudson's Bay Company controlled lands to the north and west. The canal at the American Sault removed any serious transit impediment to this westward urge. But a growing interest in the development of the region west of the Sault was bound to raise the question of Canadian access to those areas on Canadian terms. Interaction with those regions could not be allowed to be constrained by the dysfunctional portage system or the vulnerable dependence on the services of the American canal system. Military-cum-political considerations intervened to give massive weight to these very concerns.

Military Considerations and National Security

Sault Ste. Marie is a border town and the St. Mary's River is an international waterway. The War of 1812 had produced many raids and counter-raids resulting in the destruction of property and the development of a sense of armed confrontation across the border. But the American Fort Brady was not countered by the construction of a British fortification at the Sault, and after the withdrawal of the garrison from Drummond Island in 1828, the only British flag at the Sault flew at the Hudson's Bay Company establishment.

While there was considerable social interaction between the two communities during the 1820s and 1830s, the rebellions throughout Upper and Lower Canada struck a sympathetic chord with many Americans. There were no major incidents at the Sault but several minor incidents kept the international pot boiling. In February 1838 William Nourie, the Hudson's Bay Company's trader on the site, passed along to Sir George Simpson the startling rumour that a group of Canadian Métis residing in the United States had concocted a plan to occupy company property on British soil and seize what provisions the company had on hand, including several pieces of ordnance which dated back to the North West Company's presence at the Sault.[31] Nourie reported that the intervention of the commander and officers at Fort Brady had defused the situation somewhat.[32]

In April 1838 Nourie still feared an armed attack from the American shore. His efforts were directed at "keeping the Indian population on this side of the line" and moving the cannon from the post.[33] As late as July, Nourie complained that he had been unable to get the company's "arms to the Government Stores at Penetanguishine" but was moving them to Manitoulin Island to be forwarded there.[34]

In the fall of 1838 Nourie was given reason to question the continued services of his military friends at Fort Brady. September found him obliged "to keep a constant guard under arms throughout the night," but even so he was unable to prevent the burning of barns, stables and hay stacks. Certain that these fires were deliberate, he suspected the "many bad characters -- some of the self styled 'Patriots' among our opposite neighbors whose threats have been loud against the inhabitants of this side of the river."[35] These events culminated with the inhabitants of the Canadian Sault being alarmed on September 11 by an American salvo of cannon fire. On October 12 John Macauly wrote from Government House, Toronto, acknowledging Nourie's communication which had reported "the extraordinary intelligence that... several cannon shot had been fired across the water by the Garrison of Fort Brady...."[36]

Apart from these fears and minor incidents, there was no actual violence at the Sault, but the fears reappeared with the Oregon crisis of the 1840s. A War Office memorandum on the "defence of the frontier" in 1841 recommended several fortifications along the Great Lakes, including a "Fort at Falls St. Mary."[37] The fortifications were not constructed and the tensions of the 1860s occasioned a reoccurrence of the fears of American raids. In 1865 Simpson wrote, "I hope the Yankeys will keep quiet for I am in a nasty place if we get into a row with them."[38] A year later he re-echoed these fears, reporting "We are fearful of trouble with the Fenians here and I think it best to ship at once what is ready of my returns & hope they will reach you safely....There are all sorts of reports about an attack being made here."[39]

It was not until the first Riel rebellion (1869-70) and the minor international incident over the use of the American canal by Canadian forces that the Canadian Sault attracted the attention of the military. Late in 1869 Louis Riel and his Métis associates occupied Fort Garry, the strategic key to Rupert's Land, as the territories of the Hudson's Bay Company were designated. The motivation of the Métis was simple. The company was about to surrender its lands to Canada. The Métis had not been consulted and they did not want to become part of Canada without some guarantees concerning their culture and property. Consequently they established a provisional government which early in 1870 negotiated terms of entry. The Province of Manitoba was created and adequate safeguards were extended to pacify the Métis leaders. The so-called rebellion was ended by negotiations between Ottawa and Fort Garry.

Unfortunately, the rising became embroiled in the cultural conflict that was always close to the surface of Canadian life. Riel and his fellow Francophone Catholics were perceived by French Canadians as having presented legitimate demands for the protection of their faith and language. Riel became a champion of their cause. On the other hand, many Ontarians feared a plot: Quebec was using its influence to establish a French Catholic outpost in the West. Paranoia swept through Ontario and many of its leaders demanded that the rebellion be suppressed and the rebels punished.

Eventually, both sides had their way. Manitoba was created and given guarantees for language and religion; and a military expedition, consisting largely of volunteers from Ontario, was sent to the Red River

9   The *Chicora*.   (Public Archives Canada, C 48869)

10   The Red River Expedition, 1870. (Archives of Ontario, S 8243)

Valley to restore order. The expedition was placed under the command of a career officer from Britain, Colonel (later Field-Marshall) Garnet Wolseley.

After the troops had been assembled, the question was how best to get them to the distant place of conflict. One approach was to seek permission from the U.S. government to use the American rail system to get the expedition as far as Minnesota, after which the expedition could proceed north to Fort Garry by using water transport on the Red River. This was not a preferred option as it would be an admission that Canada could control the Prairies only through the offices of the United States. In any event, no American government would tolerate British troops on American soil. The alternative involved an arduous journey. Wolseley's 350 British regulars, supplemented by several hundred volunteers, would have to proceed by ship to Thunder Bay and then by land and water to central Manitoba. The expedition left central Canada in May 1870.

Supplies and arms were transported in two ships, the *Algoma* and the *Chicora*. Of course, they had to use the American canal to gain entry to Lake Superior. The *Algoma* sailed through without incident; the *Chicora* was refused passage, resulting in a great international furor. The American authorities had the right to close their canal to any foreign vessel, but mutual access to each other's canals had been an honoured custom for many years. Why did the Americans act as they did? Why did they let the *Algoma* through but not the *Chicora*?

A partial answer lies in the *Chicora's* history. In 1870 Anglo-American relations were still massively strained by disputes generated by the American Civil War. Détente was not reached until the negotiation of the Treaty of Washington in 1871. The *Chicora* had a Civil War record that rankled the Americans: she had been a blockade runner. Operating out of Bermuda, she made numerous runs into Charleston, South Carolina, and always escaped capture by the United States Navy. The blockade had been a central issue between Britain and the United States. The arrival of the *Chicora* at the American Sault was a reminder of these past tensions and frictions.

Emotions and tensions aside, it was the intervention by the American State Department to block the passage of both vessels that elevated the affair to the highest level of Anglo-American diplomacy. Eventually the *Chicora* was allowed to sail west, but Canadian vulnerability had been revealed for all to see. The communications system with the West, in the event of an emergency, was controlled by foreigners. The Toronto *Globe* returned enthusiastically to a favourite issue:

> It is quite plain...that the Dominion of Canada cannot allow the Americans to have exclusive control of the means of water communication between the great lakes of Huron and Superior, when they have shown themselves animated by so unneighbourly a spirit. They ought not to have been permitted for a single year to retain that exclusive control: and had it not been for the weakness of Sir Francis Hincks [Premier of Canada, 1851-54] all the links between the great lakes would have been held by Canada, and the American canal at the Sault would never have been built. As it is the construction of a Canadian canal is a

> necessity and ought to be commenced as soon as the necessary Parliamentary sanction can be secured.[40]

The *Globe's* views were echoed throughout the province. For example, the St. Catharines *Constitutional* offered this view:

> on this question, as upon every other matter between us and the Yankees, argument, law, reason and common sense have really nothing to do....Of all Governments within the pale of civilization, the Yankee Republic is one of the most uncertain to deal with, and is the most devoid of dignity in its administration.... It was this one idea of shameless desire to rob us of the north west that dictated this policy in preventing the *Chicora* from going through their canal.... Their conduct in this matter should teach us another lesson of not expecting favours at their hands and it also tells us that we should have a canal of our own at Sault Ste. Marie built this season.[41]

The *Chicora* went through, but many Canadians retained the notion that Canadian nationhood required a Canadian-controlled water route to the Prairies.

The *Chicora* incident was important as an international event and as an incident that convinced many Canadians a canal on the Canadian side of the Sault was essential. It also had a more local manifestation. Wolseley's troops had to be accommodated at the Sault and there was only one appropriate location — the Hudson's Bay Company establishment, which was made available to Colonel Wolseley and his troops.

The *Chicora* incident was by no means a major rupture in Canadian-American relations but it did have substantial impact on the thinking of those concerned with defence and transportation policies. They resented the fact that there was not an all-Canadian water route from Quebec City to Thunder Bay. A canal on the Canadian side at the Sault would complete that route and reduce Canadian dependency on the United States. The security of Canada's vast western empire demanded that such a canal be built. The logic did not escape the ever astute Wemyss Simpson, who reported to his superior, "I spoke to Mr. Smith about getting the patent from the Ontario Government for the land which will become very valuable if the canal is made...."[42]

The Sault as a Junction

Until the mid-19th century the Sault had been located on the periphery of settlement, on the outer fringe of the occupied area of the continent. Its location on the frontier between the two nation states imparted some importance to its location, but the Sault was not a crucial location in terms of commerce or military strategy. The volume of commodities transported by the fur companies on both sides of the rapids could be accommodated by simple portage technologies.

The second half of the century witnessed two major developments. First, the promise of substantial quantities of forest and mineral products from the areas tributary to the upper Great Lakes argued for a means of communication that facilitated the expeditious movement of bulk carriers of the new cargoes around the rapids. Second, the dramatic expansion west in both the United States and Canada resulted in the Sault becoming a location central to continental economic development. The completed and proposed canals emphasized the Sault's continued function as a corridor of east-west movement. After the extension of rail networks on both sides of the isthmus, the Sault also became the junction of rail and water transportation and the crossing point for north-south communications. Such developments elevated the Sault to the position of a major continental transport corridor.

III CONSTRUCTION OF THE CANADIAN CANAL

The Government Committed

The structure of British North America changed dramatically in 1867 when Confederation was achieved. After 1867 the concept of a canal at the Canadian Sault must be looked at within the context of the economic and transportation policies of the new Dominion. Three factors were important.

First, Canada was committed to the annexation of Rupert's Land and British Columbia. This ambition made east-west transportation systems in the upper Great Lakes much more important than had been the case even during the years of the Province of Canada. Second, Confederation was partly a product of economic nationalism and a response to protectionist and nationalist policies of the United States which emerged from the Civil War. This Canadian economic nationalism influenced the shape of all economic policy in late-19th-century Canada, and increased the prominence given to east-west lines of communication. Third, the new Federal Government of Sir John A. Macdonald favoured a policy of Canadian controlled east-west transportation systems. This became evident after the construction of the Intercolonial Railway which connected the Maritime provinces with central Canada, and after the pledge in 1871 to build a railway to British Columbia. For Macdonald, Cartier and other government leaders, that pledge included an all-Canadian route in spite of the enormous expense of building the line north of Lake Superior. Inland water communications did not attract as much attention as the railways after 1867, but they were taken seriously by the government, which had definite views on the place of the waterway between Thunder Bay and Quebec City within the Canadian economic structure.

Hector Langevin, minister of Public Works, outlined the government's policy in July 1870, only a few weeks after the *Chicora* incident. Langevin's concern was the entire water route between Thunder Bay and Quebec City, along with various feeder routes:

> The principal line of these Canals is on the St. Lawrence river, and connects the navigable waters which lie between the sea and the Upper Lakes. On this main line of navigation there are two sizes of locks, viz: those on the St. Lawrence Canals, which are 200 feet long by 45 in breadth, with a depth of 9 feet -- and those on the Welland Canal, 150 feet long by 26-1/2 in breadth, with a depth of 10-1/4 feet. Other important lines, such as the Ottawa and the Rideau, are provided with Canals and locks, varying respectively in depth and dimension.[1]

Langevin had a fixed view concerning a canal at the Canadian Sault and continued in his memorandum: "In the West our Canal System is incomplete so long as we are compelled to have resort to the United States' Government for permission to enter Lake Superior." He was equally clear about the relationship between our inland water transporta-

tion system and Canadian economic strategy:
> It may be remarked that the most important of the Canals of Canada were designed, not only with a view of affording an unobstructed passage for the staple products of its soil to the ocean, but of attracting, through the same channel, a portion of the freight passing from the West to the Atlantic, and that notwithstanding all the advantages offered by the St. Lawrence route, the bulk of the traffic referred to continues to find its way to the sea-board over the railways and Canals of the United States.

The minister recommended
> that a thorough enquiry into the subject, in all its bearing, both from a commercial and engineering stand-point, should be instituted, with a view of probing to the utmost the causes which have given rise to the state of things just alluded to, and of obtaining such reliable information as may furnish the data on which to base a scheme for the improvement of the Canal System of the Dominion, at once comprehensive and uniform, and one that shall put this country in a position to compete more successfully than heretofore with the Canals and railroads of the neighbouring Republic.

Langevin enumerated 11 specific areas of concern for his proposed commission, including "5. The construction of the Sault Ste. Marie Canal, between Lake Superior and Lake Huron." He was asking for advice, but his position on the Sault canal was revealed when he ordered new surveys at the Sault in 1870, before the commission reported.[2]

The government responded to Langevin's recommendation on November 16, 1870, by appointing a "Royal Commission to Enquire into the best means for the Improvement of the Water Communications of the Dominion and the development of the Trade with the North-Eastern Portion of North America."[3] This was Canada's fourth royal commission. The commissioners were senior and highly respected businessmen: Sir Hugh Allan, chairman (Montreal); Sir Casimir Stanislaus Gzowski (Toronto); D.D. Calvin (Kingston); George Laidlaw (Toronto); Pierre Garneau (Quebec City); W.J. Stairs (Halifax); Alexander Jardine (St. John); S.L. Shannon (Halifax). They were instructed
> to institute a thorough enquiry into the whole subject in all its bearings, both in a commercial and engineering point of view with the object of obtaining such reliable information as may furnish the data on which to base a plan for the improvement of the Canal System of the Dominion, of a comprehensive character, and such as will enable Canada to compete with success for the transit trade of the Great West....[4]

Langevin's "point 5" concerning a canal at the Sault was retained, although re-numbered as "point 6."

The work proceeded under the thoroughly competent management of Samuel Keefer, who had been appointed secretary to the canal commission. Much of the necessary data was obtained from the replies to a questionnaire sent to 2400 institutions and individuals in both Canada and the United States. Other interested persons and institutions were invited to communicate their views to the canal commission. The replies were generally in favour of the Canadian canal, as was revealed in the responses to the question "Of what interest to the commerce of the Dominion would be the construction of another Canal between Lakes Huron and Superior on the Canadian Side?"[5] Alfred Waddington, a well-informed British Columbian railway enthusiast, gave the most detailed reply. Waddington argued that although the present Lake Superior trade was small, it would grow quickly after the Sault canal was completed. The Fort Garry garrison would be provisioned from Lake Superior, and the Hudson's Bay Company would abandon its Minnesota route for an all-Canadian system. The settlement of the northwest "will soon cause an increase in the traffic on Lake Superior as altogether to change the present state of things." Waddington played heavily on the nationalistic theme, arguing that without an all-Canadian route, "our commercial connection with the North West would be at best a matter of sufferance on the part of our neighbours...." A.J. Russell, a civil servant who was an expert on the Prairies, concurred in this nationalistic view: "A Canal on the Canadian side would be of very little or no interest to the commerce of Canada; but we must consider the feeling of the Americans towards us," argued Russell, "and the probability of their shutting their own Canal, at any moment against us." F.S. Holcombe of Toronto made the same point, as did the Windsor Board of Trade. Of course, not all were enthusiastic. Kingston's Board of Trade noted that the "commerce of the Dominion is well enough served by the present Canal, if its permanent use can be guaranteed...." But overall support was strong even in the face of the huge existing facility on the American side. "Are not the lock and prism of the present American Canal the largest in America?" asked the commission. "They are the largest," replied the Ottawa Board of Trade. W.H. Smith of Owen Sound was unimpressed by this fact, however, replying, "Yes, but not too large."

The canal commission reported on February 24, 1871. It had been asked to consider 12 specific canal projects. It divided the projects into four classes of urgency: "In the first class we have placed all those works which it is for the general interest of the Dominion should be undertaken and proceeded with as fast as the means at the disposal of the Government will permit."[6] The Sault Ste. Marie canal was placed in the first class list and thereby received the full endorsement of the canal commission.

The commissioners published an engineering report on the proposed Sault canal. It noted that the site had been surveyed in 1852 and argued that there "are no engineering difficulties; on the contrary, every condition seems favorable to the construction, at a moderate expense, of a first class Canal, of the dimensions proposed for the Welland and St. Lawrence." The proposal involved a "straight cut" through St. Mary's Island. The canal was required to overcome a fall in the rapids of some 19 feet, the difference in the water level of Lakes Superior and Huron.

The 1871 proposal differed from that of 1852. No doubt Keefer wrote the 1871 report as he had prepared the 1852 proposal. He summarized the more elaborate 1852 plan, and then commented:

> It is believed...that on the more moderate scale we have suggested for the Canal System of the Dominion, it will be quite practicable to overcome the whole fall by a single lock of 18

feet lift, and thus avoid the expense of the regulating weirs which would be necessary if two locks were constructed to divide the lift. This will materially simplify the construction and operation, reduce the quantity of work to be performed, and consequently the cost of the canal, and the time of passing through it.[7]

Keefer was convinced in 1871, as he had been in 1852, that a canal at the Sault would be relatively inexpensive: "The estimated cost for a Canal and single lock — Canal 100 feet at bottom, 110 feet at surface, 13 feet deep — lock 270 x 45 x 12 including the entrance piers, and excavation to deep water, superintendant's and lock tenders' houses, is $550,000."

George Laidlaw, one of the commissioners, dissented. A Toronto grain forwarder, he became a chief promoter of efficient and cheap transportation. Described as "a visionary more than a businessman," Laidlaw promoted and was largely responsible for the successful completion of the Toronto and Nippissing Railway, the Toronto, Grey, and Bruce Railway and the Credit Valley Railway.[8] Perhaps it was this obvious commitment to the technology of rail that prompted him to oppose the Sault canal, arguing that the traffic on Lake Superior was insignificant and could be handled easily by the canal system at the American Sault.[9] Laidlaw proposed an alternative: "a shallow but rather wide barge canal" be constructed "between James Bay via the Albany over to Lake Winnipeg...."[10]

George Laidlaw's dissent notwithstanding, the report of the canal commission was a ringing endorsement of the proposal that there be a Canadian canal on the St. Mary's River. That endorsement was consistent with the Conservative government's overall transportation policy and endorsed Langevin's position. After 1871 the Conservative government was committed in principle to the Sault project.

The Long Wait

Despite this enthusiastic support, the canal was not built for many years for political and economic reasons. The report of the canal commission is dated February 24, 1871, and could hardly have been completed at a less propitious time. Sir John A. Macdonald's government, which had been extraordinarily creative since 1867, was running out of steam just when it was forced to deal with several crises.

The government survived some 32 months after the completion of the canal report. During that hectic period British Columbia was annexed, the first, if unsuccessful, contract to build a national and transcontinental railroad was negotiated, the country witnessed a major crisis over the ratification of the Treaty of Washington, the New Brunswick schools crisis made its bitter entry into national politics, Prince Edward Island was annexed, the vicious 1872 general election was fought, and the Pacific Scandal destroyed Macdonald's first Confederation government on November 5, 1873. The Sault Ste. Marie canal was not an urgent matter.

From 1873 to 1878 Canada was governed by Alexander Mackenzie's Liberal party. Mackenzie did not share Macdonald's approach to transportation policy; he did not want a national transcontinental railway with its expensive line on Lake Superior's north shore. He was willing to combine rail and water transport for communications with the West and had no objection to the use of American facilities. Mackenzie was also committed to retrenchment and governed during a period of acute depression. It would have been most surprising had his regime opted to proceed with the Sault project.

Both Macdonald and prosperity returned in 1878. Sir John and his second government turned immediately to the nationalist and expansionary policies that had defined the first Conservative administration. The national policy of protective tariffs was put in place and in 1880 a syndicate was found to build the national transcontinental railway.

During the first half of the 1880s, federal politics were dominated by the struggle to complete the C.P.R. and by the second Riel rebellion. Items not essential to the government's programme were postponed. That did not mean that the Sault canal was forgotten. Pressure to build continued and the maturing national economy focussed attention on Canada's east-west transportation system. Manitoba was founded in 1870 and during the remainder of that decade settlers flowed into the new province and the eastern portions of the Northwest Territories. Winnipeg began to mature as the Prairie's metropolitan centre and the wheat economy was established as the region's economic base. In 1876 Manitoba produced the first prairie wheat for export with some 857 bushels being dispatched to Toronto via the transportation system of the United States.[11] A more important event occurred in 1883 when James Richardson and Sons exported from Manitoba the first wheat destined for overseas. Even more significant, it was sent via the Canadian route. The partially constructed C.P.R. carried it to Thunder Bay, where it was loaded onto the ship-barge *Erin* for conveyance to ocean shipping facilities in the East.[12] The Canadian route was given added legitimacy in 1884 when the C.P.R. built a grain elevator at Port Arthur, a city destined to become one of the world's great wheat-forwarding centres. Perhaps the final economic factor of major significance was the completion in 1887 of the C.P.R.'s "Soo Line." It tied the Lake Superior section of the C.P.R. to the Michigan railway system through a branch line running through the Sault. This made the Lake Superior country of Canada accessible to entrepreneurs in both Canada and the United States regardless of their shipping preferences. They could use the rail systems of either nation or they could combine rail transport with water transport on either Lake Superior or Lake Huron.

One final source of pressure was Simon James Dawson (1820-1902). Dawson, a close associate of Allan Macdonell and his circle of westward expansionists, had been a co-leader of the Canadian government's expedition to explore the area west of Lake Superior in 1857. Dawson became an enthusiast in favour of westward expansion and was active in that cause after 1857. In 1870 he provided logistical advice to Wolseley and helped the mixed expedition get to Fort Garry. He selected northern Ontario as his home, and in 1875 was elected from Algoma to the Ontario Parliament. In 1878 he switched to the federal field and won

Algoma for the Tories in the sweep of that year. Dawson held the predominantly Conservative riding until his retirement in 1891. Dawson was tireless in his advocacy of the construction of the Sault canal, a project consistent with his long-term interest in opening the northwest as well as with his interest in serving his constituency.

The Final Decision to Build

In 1887 the government included in its financial allocations the sum of $1 million for the construction of a canal at Sault Ste. Marie. The timing was logical for the C.P.R. had been built and the major national economic policies of Macdonald's Conservative party were in place.

Simon James Dawson, Tory M.P. for Algoma, explained the government's position. He started with the Keefer plan of 1852, and stated that in 1851, 4160 tons of goods were portaged up around the falls on the St. Mary's and a further 2482 tons portaged down. Referring to the potential of the site for development, he extolled "the superiority of the Canadian side — a basin above and a basin below, and the whole length of the canal about a mile."[13] But Keefer's $500 000 canal would not accommodate increased traffic flows of 1877. The utility of the first American canal at the Sault had been clearly demonstrated by the first three decades of its operation:[14]

	No. vessels	Vessel tonnage	Freight tonnage	Passengers
1855	–	106 296	–	4 270
1860	–	403 657	–	–
1865	997	409 962		19 777
1870	1828	690 826		17 153
1875	2033	1 259 534		19 685
1880	3503	1 734 890		25 766
1885	5380	3 035 937	3 256 628	36 147

The traffic was obviously increasing and Dawson was thoroughly impressed by the massive quantity and considerable diversity of the 1886 trade figures which he read into the record:[15]

Items	Quantity	$ Valuation
Vessels	7 424	
Lockages	3 593	
Tonnage Vessels	4 219 397	
Tonnage Freight	4 527 759	
Passengers	27 088	
Tonnage Coal	1 009 999	3 534 996
Barrels Flour	1 759 365	8 796 895
Bushels Grain	19 706 858	19 312 720

Tonnage Iron	115 208	5 500 723
Barrels Salt	158 677	158 677
Tonnage Copper	38 627	7 725 400
Tonnage Iron Ore	2 087 809	7 307 331
Feet Lumber	138 688 000	2 498 384
Tonnage Silver	2 009	308 964
Tonnage Stone	9 449	94 490
Tonnage Freight	230 726	13 843 560
Total		69 080 071

Dawson noted that the traffic through the canal at the American Sault "is already up to somewhat more than half the traffic of the Suez Canal which in 1885 was 8,985,411 tons." Not only should the Sault be developed for its commercial potential, its power should also be exploited. That was the American strategy: "Now they are not only building a new canal on the American side, but are cutting a canal three miles long behind the ship canal in order to avail themselves of the water power." Dawson argued that a canal at the Sault "will be the means of building up a city in a very short time at Sault Ste. Marie, and of drawing traffic to our great Canadian Pacific Railway." He concluded by referring to the *Chicora* incident and echoing the now familiar nationalist appeal. He reminded his listeners that in 1870 "the Americans had shut their canal in order to cut off all intercourse between this section of the country and Lake Superior." The appeal to the *Chicora* incident is interesting because any military problems concerning communications with the Prairies had been solved by the completion of the C.P.R. Troops to crush the 1885 rising had been rushed west by rail. The national security argument for the Sault canal made little sense.

Although the government got its one-million-dollar appropriation for the canal, almost a year lapsed before an order-in-council was issued authorizing construction. That order, dated May 2, 1888,[16] sparked the most extended parliamentary debate on the issue. It is not surprising that in the highly partisan politics of the 1880s the issue was politicized. The debate took place on May 15 and May 19, 1888. The chief government spokesman was Sir Charles Tupper, minister of Finance, and not J.H. Pope, the somewhat undistinguished minister of Railways and Canals. Defending the Sault canal was Tupper's final major service as minister of Finance.

The debate was sparked by the government's need for a revote of the money granted by Parliament in 1887. Actually, $2350 had been spent before December 1887 to "verify the surveys" and Tupper was really asking for the balance of $997 650.[17] Tupper argued, wrongly, that the tonnage through the American canal was greater than the tonnage through the Suez. He then stated succinctly what might be described as the "multiplier effect" case for the canal, which in essence argued that increased western settlement would increase traffic which in turn would benefit all of Canada. Tupper was answered by J.F. Lister, a Liberal M.P. for Lambton West, who argued against the construction of the canal. He had been assured by a variety of shippers that there was no need whatsoever for such a canal. The American canal provided access to both American and Canadian shipping at no charge and this competition would prevent the proposed Canadian canal from raising revenue. Lister dismissed the defence argument because a Canadian canal could not be protected in the event of war. Furthermore, he argued that the proposed canal made no economic sense and that the government's estimates were hopeless, predicting that the work "will cost at least double what they are asking." More importantly, "they are undertaking a work when the bill is already filled by the American Government and when we have a right to use the American canal...." and it was unlikely there would be any American interruption of Canadian traffic through their canal: "So long as we own the Welland Canal we know we shall have the right to extract terms from the United States Government that will permit our vessels to pass through the Sault Ste. Marie Canal, and the second canal that will be constructed hereafter. The Welland Canal is as important to them as the Sault Ste. Marie Canal is important to us...." Actually, concluded Lister in the partisan terms that defined politics in the 1880s, there was "no justification whatever for undertaking this work, except for the purpose of securing a seat for the Hon. member for Algoma."

Dawson leapt back into the fray, using American construction and proposed construction to justify the Canadian canal. Lister, Dawson charged, had made the case for a Canadian canal. Why would the Americans plan a second (and possibly a third) canal unless traffic was increasing rapidly? This increase in trade justified a Canadian canal. Then came what might have been the ultimate reason for the Canadian decision: "In order to keep up with the Americans we must build a canal on our side of the river."

Dawson, Tupper and Lister covered virtually all of the arguments considered in making the final decision. The Liberals exploited one weakness in the Tory case. Two high-powered Liberals, Sir Louis Henry Davies (former premier of Prince Edward Island) and A.J. Jones (former minister of Militia) focussed on the cost argument. They ridiculed the one-million-dollar estimate, Jones suggesting the cost might be four to five times that sum. Davies and Jones were Liberal partisans; they were also maritimers who harboured profound suspicions about the improvement of inland waterways. Most maritimers assumed that such improvements would weaken the deep-sea ports on the Atlantic. Nonetheless, their points struck home.

The debate continued in the form of interminable partisan nit-picking. Finally, Charles Wesley Weldon, a ranking Liberal from Saint John, New Brunswick, scored heavily:

> Last year, the vote was granted on the statement of the hon. member for Algoma, and what passed in concurrence when the hon. Minister of Finance was asked to explain the vote? The following is the report in *Hansard:* "Sir RICHARD CARTWRIGHT. Have tenders been asked for this work? "Sir CHARLES TUPPER. Not Yet. "Sir RICHARD CARTWRIGHT. Have any reports of engineers been received? "Sir CHARLES TUPPER, Elaborate reports were made some time ago, on two occasions, and Mr. Page is now considering the whole question. Very full plans and estimates are in the department, and

Mr. Page is considering the whole matter." If Mr. Page has been considering the whole thing since last Session, we have the right when asked to vote this amount to know what his views are, and his estimate of the probable cost.[18]

Sir Charles was sufficiently provoked to reconsider his figures. He returned to the House of Commons on May 19 and raised the Sault canal estimate to $2.8 million. He got his money: the way was now cleared to begin the actual construction of a Canadian ship canal at Sault Ste. Marie.

Awarding the Contracts

The single lock built by Canada at the Sault was, on completion, the largest such structure in North America. This was a very large and complicated project in terms of the evolution of the design, the decisions behind the awarding of tenders and contracts, the immensity of the actual construction required, the provision of supplies and labour to the site and the whole question of federal, provincial and municipal interaction on the project. Moreover, nothing about the project was straightforward.

The government was ready to proceed in the spring of 1888, but was still not certain it knew what it wanted. In a report prepared by John Page for J.H. Pope, the minister of Railways and Canals, Page noted that Pope had "intimated a few days ago that it had been decided to proceed with the construction of a Canal at Sault St. Marie."[19] He continued: "there are several questions connected with the matter that must be determined before the design can be intelligently prepared." Page's points were fundamental: first, "The depth that the canal is intended to be at the lowest known stages of water above and below the rapid"; second, "What kind of Lock it is contemplated to build that is to say shall it be a long large structure suited to receive a number of vessels at the same time, or shall two Locks be built and placed side by side each capable of passing the largest class of vessels now engaged or likely to be engaged on that line of navigation." Page's concerns were resolved within the Department of Railways and Canals, but not definitively. As the work progressed the government had second thoughts that required substantial changes in the initial design and structure of the work.

The government's general position was stated in the Railways and Canal's 1889 annual report:

> This canal is intended to be constructed on the Canadian side of the River St. Mary between Lakes Huron and Superior, being formed through St. Mary's island on the north side of the rapids. At ordinary stages of the river water there is a difference of 18 feet in the levels of the water above and below this island. The distance across the island is about two thirds of a mile. The canal will have a mean width of 150 feet and a depth of 18 feet below the lowest known water line of that part of the river. The

NOTICE

TO CONTRACTORS.

Plans and Specifications of the work to be done at the

UPPER and LOWER ENTRANCES OF THE

Sault Ste. Marie Canal

Can be seen at this office and at Sault Ste. Marie, Ont., on and after

FRIDAY, THE 23rd DAY OF NOVEMBER, NEXT,

Where forms of tender may be obtained.

Tenders for the execution of the works, addressed to the undersigned, and endorsed, "Tender for Sault Ste. Marie Canal," will be received until the arrival of the eastern and western mails on

FRIDAY, THE 7th DAY OF DECEMBER, NEXT.

By order.

A. P. BRADLEY,
Secretary.

Department of Railways and Canals,
Ottawa, 26th Oct., 1888.

11 "Notice to Contractors," 1888. (SSMCC)

difference in level will be overcome by one lock 600 feet in length and 85 feet in width, having guard gates at both ends facing in opposite directions. These gates are to be worked by 'Hydraulic power,' the water being admitted or withdrawn at the floor of the lock....The contracts require the whole to be completed for use in May, 1892. The canal will be crossed by the railway leading to the Sault St. Marie Railway Bridge recently built. The crossing is to be made by a swing bridge of sufficient capacity to span the canal.[20]

The overall project was to be divided into three sections, the main work comprising the production of navigable channels, navigation beacons and piers at the eastern and western entrances, together with the key element, the canal prism and lock unit. Other smaller, though no less essential, projects were tendered as the construction progressed, of which the following is only a small sample:

1893   "the Construction of Steel Pipes and the Valves."[21]
1893   "the Construction of Valves, Gratings, etc., to be placed in the Lock now under Construction."[22]
1894   "machinery for operating Gates and Valves, including all bed-plates, pulleys, cover-plates, crabs, attaching cables etc., and whatever may be required for the full completion of the operating machinery."[23]

The Department of Railways and Canals provided detailed specifications for every minute phase of the construction process. Those for the machinery for operating the gates and valves, for example, consist of 36 separate points that require 12 sheets of foolscap to print. Item 6 alone, "Brass or gun-metal," demonstrates the degree of specificity typically provided for the contractors:

> The Brass or gun-metal comprising the nuts or female screws through which the large screws operate and also all brass journal linings and collar friction plates and wherever Brass is used throughout the machinery, crabs etc., shall be composed of gun-metal, which shall be of a strong and durable quality more especially the portions forming the large nuts or female screws being of the proportions, viz: 16 parts copper and 3 parts tin or of such other proportions as may be approved by the Chief Engineer of Railways and Canals, the castings must be sound, free from flaws, turned to a fine finish and so as to ensure solid bearings both on the outside and inside of the metal and to the dimensions shown on the drawings, giving such allowance for play for shafts as is customary in practice or as may be approved by the Chief Engineer of Railway and Canals.[24]

The specifications for the lock gates for the lift lock consist of 43 points requiring eight pages of small print. It took 19 pages of small print to provide the requirements for the superintendent's residence. Essential though they were to the quality control of the project and the proper financial management and supervision of the agencies involved, such specifications make for dull reading. The engineering drawings produced to guide the contractors are not only a testament to the penmanship and technical skills of the draughtsmen who rendered them, but also they merit recognition as another dimension of the industry, concern for detail, and competence of the late-Victorian industrial era.

Of all the elements of the project, it was "Section 2," the construction of the canal prism and lock, that attracted the most attention. Tenders were received from all over Ontario and Quebec and even the United States:[25]

	Company	Location	Bid
1.	Goodwin	Ottawa	$1 163 692
2.	McTergue, Dwyder, Ray, Commee	Port Arthur	$1 225 990
3.	Ryan, Haney	Toronto, Brockville and Watertown, N.Y.	$1 282 567
4.	Nelson, Caroll, Shields	St. Catherines	$1 320 282
5.	Raynor, Belden	Syracuse, N.Y.	$1 450 806
6.	McArthur Bros	Chicago, Ill.	$1 385 650
7.	Murphy, Gray	Quebec City	$1 525 155
8.	Murray, Cleveland	St. Catherines	$1 547 132
9.	McDonald, Aylmer	Toronto	$1 604 511
10.	Ross, Holt, Mackenzie	Sherbrooke	$1 656 524
11.	Reid	Montreal	$1 805 120
12.	Larkin, Connelly	St. Catherines and Quebec City	$1 912 686
13.	MacLellan, MacLellan	Toronto	$1 954 165
14.	Parry, Macdonald, McCallum, LaFinneure	Ottawa	$1 990 144

After due consideration, cabinet decided in favour of Ryan and Haney, a firm known as Hugh Ryan and Company. The successful bid was $1 282 567. The two lower tenders, those of Goodwin ($1 163 692) and McTergue, Dwyer, Ray and Commee ($1 225 990) were rejected because their totals were less than the purchase and delivery costs of materials.[26] Hugh Ryan and Co. were also awarded the contract for "Section 1" while another company, Allan and Fleming, were contracted for the work on "Section 3." But it was the Brockville-based Hugh Ryan and Company that was given the responsibility for the bulk of the work at the Sault. Ryan was an experienced contractor as was his chief partner, Michael John Haney, who had joined with Ryan in 1886 to build the railway line from Winnipeg to West Lynn that later became part of the Canadian National Railway system. Haney was to move on from the Sault Ste. Marie ship canal project to a career as a major figure in other Canadian construction and business enterprises. For Ryan, however, the Sault project was to prove to be his last major venture.

Even after the contracts were awarded there was some uncertainty regarding the specifications of the crucial element of the project, "Section 2." Interested elements in both Montreal's and Toronto's business communities wanted an even bigger canal that could accommodate the larger ships that were being planned then. Montreal spoke through the Canadian Pacific Railway. As early as August 16, 1890, the secretary of the C.P.R. responded to some concerns expressed by company president Sir William Cornelius Van Horne, which were related to the basic dimensions of the Canadian and American locks. The firm's secretary noted that current plans called for a Canadian canal with a draught on the mitre sill of 16 feet 8 inches, 600 feet long and 85 feet wide. The new American canal, however, was to have a draught of 18 feet 11 inches and would measure 800 feet by 100 feet.[27] Van Horne wrote to Sir John A. Macdonald the very next day to recommend that the Canadian lock be at least as deep as the new American one which was to be more than 2 feet deeper in draught than its Canadian rival: his chief argument related to naval warfare. He argued that Canada must be able to get ironclads onto Lake Superior that were as large as any that the Americans might place there.[28] Van Horne pointed out that enlargement at this stage would be relatively cheap compared with enlargement after the completion of the project.

Macdonald turned the matter over to W.G. Thompson, who prepared a careful report. No conclusion, he argued, is possible without a careful examination of the relationship of the navigable channel in the St. Mary's River to the system above and below. Thompson explained that the situation above the main canal site, between the Sault and Lake Superior, was relatively straightforward. There the "navigable channel is so clearly defined by nature that no question can arise as to its location." It is a "fine channel" with sufficient depth of water "and being the international boundary is free to all vessels alike."[29] Below the Sault, the situation was more complicated, however, and Thompson underscored the implications for the Canadian system of certain American developments. There, the Americans were "deepening the 'Hay Lake' Channel to a depth corresponding with that of the new lock now under construction by them at Sault Ste. Marie." Not only was this all-American channel deeper, it was also 8 miles shorter than the "boundary line channel." For Canada to match this by an equivalent all-Canadian route would require the dredging of the latter which would still suffer from the disadvantage of being the longer route. To Thompson, the issue was a crucial one which had ramifications beyond the local area alone. He argued that before deepening the Sault Ste. Marie canal, the government should decide whether or not it intended to continue with the present policy of a 16-foot system of navigation from Port Colborne upwards. If so, the improvements suggested for the Sault would be futile.

Thompson pointed out that Canadian-American relations were cordial and that the 19-foot U.S. lock was available to Canadian vessels. In the event of a deterioration of relations with the Americans, the trade through the Sault could be transferred to the C.P.R. Hence, concluded Thompson, the depth of the Sault canal was a secondary consideration. Also, if the lock were to be deepened at this stage, he noted, it would necessitate the lowering of the prism and require the destruction of the side walls. These changes would mean breaking the contract with Ryan and Haney, "so that the cost of making such alterations can hardly be arrived at in the usual way, the door being opened for 'damages.'" Thompson concluded: "If asked an opinion in the matter, I would say that the circumstances do not warrant the adoption of a course that would establish such a dangerous precedent as the breaking of an important government contract."

There the matter rested for the winter, but Van Horne was back in the spring. This time he was less oblique. The military argument was gone, and Van Horne's position had been strengthened. The government had won a tough election, requiring the aid of the C.P.R.

Van Horne explained vessels carrying 3000 tons would certainly be built in the next 6 months, necessitating modification of the canal to accommodate such vessels: "it will cost infinitely less to make the lock deep enough now than to make the correction in the future...."[30] Toronto's business community spoke through the marine section of the city's Board of Trade. It wanted 20 feet "of water in the sill..., being the same as that of the new lock under construction on the U.S. side at Sault Ste. Marie."[31]

T. Trudeau, chief engineer of canals, was sympathetic with the idea of enlargement, urging even more expansion than either Van Horne or the Toronto Board of Trade. Trudeau recommended 19 feet of water on the sills and a lock chamber measuring 650 feet (rather than 600) by 100 feet (rather than 85) which would accommodate four large ships simultaneously.[32] Macdonald was well aware he owed his narrow victory in March 1891 to the C.P.R. and other great business concerns: the government capitulated and ordered that a report be prepared for the minister of Railways and Canals concerning suggested changes in the work in progress at Sault Ste. Marie.[33] On May 29, 1891, the Department of Railways and Canals made six suggestions to Hugh Ryan and Co.: 1. widen the lock chamber to 100 feet; 2. increase the length to 650 feet between the hollow quoins.; 3. deepen the lock to give 2 feet, 9 inches more water on the mitre sills than originally specified; 4. eliminate the circular recesses in the lock walls intended for the reception of the filling and emptying valves, thereby decreasing the length of wall to be constructed by 40 feet; 5. increase the dimensions of the filling and emptying culvert pits; 6. complete the additional work in points 1-5 for $219 000.[34]

Ryan accepted and the reorganized agreement was implemented in June.[35] This was only the beginning of modifications to the original proposal. Later in 1891 a further supplementary contract provided for an entrance to the canal that was in line with the redesigned lock chamber, thus affording approaching vessels with "the advantages of a straight entrance and exit into and out of the chamber...."[36]

The ever optimistic T. Trudeau retained his interest in building a bigger canal. When Sir Richard Cartwright suggested "the expediency of increasing the width of the gate opening to the full width of the proposed lock, namely 100 feet...," Trudeau prepared six different designs. The sixth scheme was described as follows:

> I may, however, observe that a still further scheme has been suggested by which provision would be made for one Lake vessel 320 feet long and two Welland Canal type vessels 235 feet long. There are no engineering objections to such a scheme, which may be indicated as follows: Length of Chamber (between lock gates) 900 feet; Breadth 60 feet; Breadth of gate openings 60 feet; Depth of water over the mitre sills 19 feet; Number of vessels to be passed at one lockage 3; Estimated cost $1,600,000.[37]

Mackenzie Bowell, the acting minister of Railways and Canals, opted for scheme number 6, implemented by order-in-council, April 1, 1892. This major change in design provoked a discussion in the House of Commons. Sir Richard Cartwright, a former minister of Finance, noted that the canal's original width was to have been 85 feet; it had then been increased to 100 feet only to be decreased later to 60 feet. He suggested 60 feet wide was insufficient for a lock 900 feet long and asked "What were the reasons for the changes?" J.G. Haggart, minister of Railways and Canals for less than 4 months, replied for the government: "Originally it was intended that vessels should lie side by side in the lock as they do on the American side, but afterwards it was thought that it would delay locking and that the present dimensions are sufficient to accommodate all the vessels."[38]

Ryan agreed to complete Trudeau's "scheme 6" by December 30, 1894. Even that was to be modified and late in 1892 the contractors agreed to produce an extra 4 feet of depth in the prism and a new completion target of Dominion Day 1894. Of course, the series of changes in specifications and deadlines complicated the contractors' work and gave them substantial political leverage with the government. Hugh Ryan claimed that these changes plus several directives from the department increased his costs.[39] Even the new Liberal government of Wilfrid Laurier agreed that Ryan had a case. The matter was submitted to a one-man arbitration in 1897. The choice of arbitrator indicated the extent to which the new minister of Railways and Canals, A.G. Blair, was sympathetic to Ryan. Walter Shanly was a civil engineer closely connected with the Conservative party of Sir John A. Macdonald. He had been general manager of the Grand Trunk Railway, 1858-62, and Conservative M.P. for Grenville South, 1863-72 and 1885-91. The 82-year-old Shanly conducted an exhaustive enquiry. His report was 1127 pages and was accompanied by several dozen technical appendices.[40] Hugh Ryan and Co. received an award of $211 505.

Progress of Construction

The work at the Sault was organized into three units, as mentioned earlier. The contractors dealt with each unit separately, as did the public service in its monitoring and reporting on the progress of the huge construction project. Before the construction of the canal is described, it is appropriate to detail the specifications of these three sections:

Section one: This involved the preparation of a 5300-foot navigable channel in the river below St. Mary's Island. The channel is to be 18-1/2 feet deep at lowest water and 250 feet wide at bottom. This section also involved the construction of entrance piers and a beacon. Hugh Ryan and Co. were awarded this contract.

Section two: This was the main contract and was also awarded to Hugh Ryan and Co. This involved the construction, on St. Mary's Island, of the lock pit and prism of the canal. Side walls and puddle trenches were to be completed and a lock constructed, together with its masonry and gates. The initial dimensions of the prism were to be 145 feet at bottom width and 18 feet deep. The initial dimensions of the lock chamber were 600 feet long between the hollow quoins, 85 feet mean width, 60-foot-wide entrances, and 16-1/4 feet depth of water over the mitre sills. Under normal conditions, the lock's lift was to be 18 feet. Emptying and filling culverts were to be constructed and, according to the 1889 specifications, the system was to be powered by a hydraulic (not electric) system.

Section three: Allan and Fleming were awarded the contract for the upper or western section and they were to construct 9300 feet of navigable channel above St. Mary's Island. It was to be 250 feet wide and 18 feet deep at lowest water. A beacon and entrance piers were also required.

Although these specifications were to be modified repeatedly, they do explain the task at hand as the contractors began work in 1890.

Preliminary Stage: Preparing the Site, 1887-90

The report of federal government expenditures at the Sault reveal that real work on the canal construction began in 1890. No funds were used in 1887; only $8145 was spent in 1888; in 1889 the amount increased to $34 018.95; in 1890 the Department of Railways and Canals reported that $176 508 had been applied to the canal and lock at the Sault.[41]

An analysis of how money was spent in 1888-89 underscores the point that before the summer of 1889 the work involved preparing the site for construction. Of a total expenditure of $34 018.95, Allan and Fleming spent $10 700, while Hugh Ryan and Co. used a mere $4900. The remainder was spent directly by the bureaucracy: $1800 covered the expenses of the resident engineer and his assistant; rodmen and axemen involved in the survey received $4906; the general expenses account absorbed $8667.30; only two other costs exceeded $200 -- travel expenses amounted to $230.95 and office rent at the Sault totalled $260.[42]

Expenditures during the fall of 1889 reveal that actual construction was now underway, with excavations for both the canal and lock being in progress:[43]

Hugh Ryan and Co. Estimate, October 6-31, 1889

Chopping, clearing and grubbing	1 050.00
Earth excavation in prism of canal	5 848.25
Rock excavation in prism of canal	1 168.20
Earth excavation in lock pit	7 340.00
Rock excavation in lock pit	11 912.05
Underwater lock pit (on bulk sum of $20 000)	1 200.00
	28 518.50

Allan & Fleming Estimate, October 6-31, 1889

Excavation, dredging and deepening bottom	65 523.35

"Chopping and clearing," "excavation," "underwatering" fail to invoke an image of the actual work involved. Contemporary photographs help to fill out this picture. The site was a muddy irregular scar of giant proportions cut across St. Mary's Island and the scene of considerable activity as the scores of men went about their varied specialized tasks. Only the sounds are missing. The air must have been full of the detonations of the blasting operations, the cries of men and animals engaged in the hauling and carting of the supplies, and the steady beat of the pumps "unwatering" the deep excavation for the lock.

Land ownership and the maintenance of law and order on the site were important issues. The latter was relatively simple, requiring only a discussion concerning jurisdiction. The Department of Justice was asked to decide whether Ontario or Canada was responsible for the maintenance of law and order on the site.[44] The infusion of large numbers of workers into the area was a potential threat to the peace and security of the host community.

The land issue was much more complicated, largely because the construction of the canal increased the value of neighbouring properties and attracted the attention of several entrepreneurs to the potential for industrial and power development. In July 1888 Arthur Stephen attempted to acquire a portion of St. Mary's Island.[45] The Department of the Interior sought information concerning the needs for lands in the area in connection with the proposed canal.[46] They were advised that all of St. Mary's Island should be reserved along with the islands to the east and downstream of it and the islands between St. Mary's and the river.[47] But the pressure on adjacent lands continued and as is often the case in Canadian affairs, some jurisdictional problems had to be faced. Thus, in 1889 John Page, then chief engineer of canals, informed the secretary of the Department of Railways and Canals that the Province of Ontario planned to sell at auction several parcels of land located adjacent to the canal. Page feared that private possession of those lands could interfere with canal construction, and he recommended that Ontario be informed that these lands were needed for canal purposes.[48] Four days later, Page learned from Toronto that the lands had been withdrawn from sale.[49] It was not until 1892 that legal title to St. Mary's was vested in the Department of Railways and Canals and the question of the ownership and control of the adjoining lands was to be a point of issue for the next two decades.[50]

The First Report Year, 1889-90: Work Begins

Hugh Ryan and Co. were responsible for construction in the portions of the original contract specified as "Section 1," the lower entrance to the canal, and "Section 2," the lock pit and prism of the actual canal itself. The lower entrance involved the dredging and excavation necessary for a navigable channel in St. Mary's River below the foot of St. Mary Island. During the first report year ending June 30, 1890, almost all of the work completed on "Section 1" was dredging and some 14 438 cubic yards of material were removed. Contractual "Section 2" was the primary site, however, requiring a 3500-foot excavation along the length of St. Mary Island. Here Hugh Ryan and Co. were engaged in the excavation of the lock pit and canal prism, construction of side walls and puddle trenches and "the construction of a lock and masonry for a guard gate."[51] When the contractors began work, they were using the specifications included in the original contract. Compressed air for the construction site was provided by a hydroelectrical power plant constructed by the contractors near the site, although it had to be supplemented by steam power because of fluctuations in the water levels of the St. Mary's River and "the difficulties arising from a winter temperature far below zero." During this first report year substantial progress was also made on "Section 2," and on several fronts. First, the excavation of earth from the lock pit had been completed and 60% of rock removed. As would be expected, major attention had been directed to the prism of the canal and it was reported that "rock has been stripped for about 700 feet above the lock pit, and for the remaining length of the section the surface boulders have been removed, and the excavation got into shape to admit of systematic work." Early in 1890 the contractors began to quarry at the Anderden quarries in Essex county, stockpiling the stone that had been used successfully on the Welland. The statistical statements in the report reveal much about the quantity and focus of the effort that Hugh Ryan and Co. were able to give to "Section 2" during the first year of construction. Thirty acres of land were "grubbed" and cleared; 2057 cubic yards of earth and 74 of rock were excavated in the side trenches; 29 045 cubic yards of earth and 1349 of rock were removed from the prism; 35 240 cubic yards of earth and 43 886 of rock were excavated from the lock pit.

Hugh Ryan and Co. had problems during their first full year of operation for large-scale construction operations are always fraught with uncertainty. Ryan proposed a change involving trenching. By the end of 1889 he had excavated the lock pit to the required depth and suggested to the department that he had sufficient power to keep the lock dry without puddle trenches around it; the nature of the rock formation in the excavation resulted in substantially less leakage than anticipated.[52] In the spring of 1890, Ryan ran into problems with the stone from the Anderden quarries. The specific thickness of stone required was 18-24 inches and the beds being quarried at Anderden were running at 20, 22 and 24 inches thick. Ryan suggested that substantial savings could be realized in working and dressing stone if thicker stone with a minimum thickness of 22 inches were to be used.[53] Ryan also had some financial complications. He had given the government more than $28 000 as a security deposit for the completion of the canal. Ryan suggested that the government accept real estate — residential and commercial property — in Brockville in lieu of the cash. The government agreed, but only to the amount of $24 000.[54]

Allan and Fleming were responsible for the upper entrance at the Lake Superior approach to the canal. This involved a 9300-foot-long channel from the head of St. Mary's Island to the navigable waters above. The channel was to be 250 feet wide and 18 feet deep "at the lowest recorded water surface above the falls" and provided with a beacon and entrance piers to the lock.[55] During their first report year, 110 511 cubic yards of material were dredged out of the channel.

Like Hugh Ryan and Co., Allan and Fleming faced some initial financial problems. Just before Christmas 1890 the firm explained to the ministry that it had made substantial progress in excavation (81 020 yards in one working season) and that it would have to stockpile "timber and other material" for the pier construction in the spring. Consequently, Allan and Fleming asked for $10 000 from the "security and drawback" fund.[56] John Page, chief engineer of Canada for the Dominion, judged their request "a little premature," but suggested that "they might be allowed half the retained sum of $4,000."[57] Unfortunately, John Lorn McDougall, Canada's auditor-general, declared the request to be legally questionable and there the matter rested for the time being.[58]

Even the civil service engineers had problems. They were obliged to live on or near the site and required to observe and inspect every phase of the work. John Page was on the site in 1889 and he observed that the rooms he and his colleagues occupied were suitable for the summer but inappropriate for winter occupation. It was logical to relocate the offices on the canal site where they could also accommodate the canal's repair staff.[59] The order-in-council authorizing the engineer's office was passed August 7, 1890.[60]

The Second Report Year, 1890-91: Work Continues

Substantial progress was made by Hugh Ryan and Co. on both sections of their contract during the second report year ending July 30, 1891. The excavation of the lower entrance was roughly half completed. Work was under way on construction of the north landing pier: cribs were in the process of being sunk. Some sense of the gains being made may be gathered from the quantities of excavations made and materials being used: 80 901 cubic yards of material were dredged out of the channel; 2245 linear feet of timber were used in cribs and as bottom ties; modest quantities of material were absorbed as binding pieces and blocks under heads of ties; 485 pounds of wrought-iron bolts were needed and 40 cubic yards of stone used to fill cribs.[61]

Work on the main section also progressed, complicated by a major change in design. The canal had to be deepened to 19 feet but even this was less than clear: "This depth [i.e. 19 feet], though calculated on a different basis (extreme low instead of 'mean' water level), is intended to be the equivalent of the depth, 21 feet, of the new American lock now under construction."[62] This modification delayed the completion date to May 1, 1893.

By June 30, 1891, a third of the canal prism had been excavated, the enlargement begun, and the contractors could report considerable progress. Earth and rock excavation in the canal prism totalled 69 099 cubic yards, and earth and rock excavation in the lock pit, 40 419 cubic yards. This total of 109 518 excavated cubic yards represented a major advance in the construction. During the same report year, 3760 cubic yards of rough stone for masonry work were delivered to the site.[63]

Ryan informed the department that the whole lock pit had been excavated by Thanksgiving and that the amount of water in the excavation was imperceptible. Leakage was being pumped out without difficulty. Ryan reiterated the point that puddle trenches were unnecessary.[64] W.G. Thompson agreed with Ryan and suggested that in lieu of trenching, a cement wall be constructed. The department agreed but held back on Thompson's wall. It was explained to Ryan that the puddle trenches had been designed both to exclude water from the excavation during construction and to prevent water from escaping after the completion of the canal. Dispensing with the puddle trenches was acceptable to the department but two conditions were imposed: 1. responsibility for unwatering the works was to remain as if puddle trenches had been constructed; 2. if necessary, a cement wall must be constructed.

Vast quantities of stone were being quarried at Anderden. This continued to concern Ryan because it was removed from the quarry in 28-inch thicknesses, a measurement not yet accepted by the department. In August 1890 alone, 1000 tons of stone were delivered to the site.

Ryan was unhappy with the amount of work accomplished during the summer of 1890. The weather had been unusually wet. More important, there had been some difficulties with the work force of 150 men. The summer had witnessed labour troubles and a strike. This is an intriguing issue that receives only a passing mention in the evidence.[65] There was a second strike with 400 stone cutters in 1894. These events also underscore that little is known about the social history of the men who built the Sault canal; however, conditions must have been unpleasant. Blackflies abounded during the early summer. Housing was primitive and in short supply in what was still a frontier district. Recreational facilities were virtually non-existent and the winters were long and cold. Apart from these conditions, the work itself was potentially dangerous although it seems that only one worker was killed during the construction period.[66] The construction force was large. On site there were usually about 400 men working at the excavation and construction proper, plus 200 stone masons. Another 600 men were employed at the Anderden quarries cutting the stone for facing the locks, others working at quarries on Manitoulin Island to produce the backing stone. The Sault site also hosted "53 teams of horses and 10 miles of railroad track." All of this was overseen by the site boss,

Michael Haney, whose vigilance was assured by his partnership in the Ryan company.

Progress was also maintained at the western or "upper" approach to the lock. Allan and Fleming had completed three quarters of the dredging and the beacon was "practically completed."[67] Only the quantities of excavation and of materials used demonstrate the scale of the project. From the channel an impressive 194 814 cubic yards of material was dredged; 6392 cubic feet of rock elm was used in the beacon; 10 760 linear feet of cross ties were consumed; 2018 cubic yards of stone filling were required; 21 700 board feet of 5-inch sheeting and pine planks were used; 20 400 pounds of wrought-iron bolts, straps and pressed spiked were required by the construction crews; smaller quantities of wooden blocks and planks were also needed.

Allan and Fleming continued to have financial problems. Late in 1890 the firm tackled the question of the drawback fund once again, pointing out to the department that two-thirds of the excavation had been completed and that the beacon was almost finished. The firm asked for the full amount of the drawback, $14 583, and this time the department was receptive.[68] Thompson was prepared to give Allan and Fleming $14 000 of the drawback funds requested, withholding $2000 for unforeseen circumstances.

The Third Report Year, 1891-92: Solid Achievement

By the end of the third report period on June 30, 1892, total expenditures had grown to $1 182 762.43. There were signs that these would increase substantially beyond the earlier estimates. One reason was yet another alteration of the original design, the lock now being required to be 900 feet long. The extra work also resulted in an extension of the completion date to July 1, 1894.

The project was far from completion although the work continued to progress. At the lower approach the construction of the beacon was abandoned "in the interests of navigation" and the funds budgetted for its construction were directed to "increasing the length of the north pier."[69] Improving the navigability of the channel there required not only dredging, but also drilling and blasting to attain the depths required. There was no sign of any slackening in the intensity of the work there. Compared with selected figures for preceding years, quantities of materials used or excavated in 1891-92 illustrate how the work was both progressing and changing. In 1890-91, 80 901 cubic yards of material were dredged from the channel; in 1891-92 only 41 205 cubic yards were removed, the channel being now close to completion. However, 67 935 linear feet of timber were used in cribs, ties, stringers, binding pieces, and cap pieces in 1891-92. This had amounted to only 2317 linear feet in the previous year; clearly pier construction was the new focal point of concern. Substantial amounts of other materials were used for purposes of construction in 1891-92: 1226 blocks under ties; 31 418 pounds of wrought-iron bolts, pressed spikes and iron bolts; 7094 cubic yards of stone filling; and huge quantities of timber in a variety of forms and dimensions.[70] The dredging continued and would do so for several years after completion, but the piers at the eastern entrance were close to completion and ready for the operation of the lock.

Progress on the construction of the lock had been affected by the April 1, 1892, order-in-council, which had increased the length of the lock to 900 feet. Although this order had also extended Hugh Ryan's completion date to December 31, 1894, it had excluded certain key items of the works. It was recognized that "the filling culverts in the bottom of the lock, the gates, valves and operating machinery" were to be built and installed after December 31, 1894.[71] Despite the extra work demanded at short notice, by June 30, 1892, Ryan could report that "the excavation of the lock pit for the 900 feet lock, was near completion, and the delivery and preparation of materials for the lock, had progressed fairly...." An amazing amount of earth and rock was excavated. In all, 178 111 cubic yards of material were removed from the prism, the lock pit and the culvert pits. Vast quantities of supplies continued to arrive at the site: 14 304 cubic yards of dressed or rough stone was delivered; 3771 cubic yards of sand and 5300 barrels of cement arrived; 9153 cubic feet of pine and oak timber, and 85 599 board feet of planks for culverts and the mitre sill platform were unloaded at Sault Ste. Marie.

The contractors at the upper end of the project, Allan and Fleming, continued to have financial trouble during the third year of their contract. They requested $16 300 from their drawback of $23 981.[72] This time, however, their perennial appeal was not so favourably received. Thompson reported that some of the firm's dredging subcontracts were in trouble and that the department was finding it difficult to get the dredging to the required depth. Thompson recommended no advance.[73] Nonetheless, the work went on at the troubled northern "Section 3." A further 20 125 cubic yards of earth and rock were removed from the channel; 71 241 linear feet of timber was used along with 1708 cubic feet of timber and 1793 blocks for ties and 2071 board feet of sheeting and planking; bolts, straps and spikes totalled 23 551 pounds; 6033 cubic yards of stone filling were used.[74] The quantities reported reveal the prodigious amount of dredging required at the western and upper approach and indicate the problems encountered by the contractors. The beacon, however, was close to completion.

The Fourth Report Year, 1892-93: The Lock Takes Shape

By the time of the fourth report (June 30, 1893), the canal and lock were taking shape. During that year of construction, $589 801.25 had been spent, bringing the total expenditure to that point to $1 475 344.45.[75] Clearly, a substantial physical plant had come into existence at the site by the end of the fourth year of activity. Departmental officers were both optimistic and enthusiastic: "The masonry of the lock has been all executed, and the remainder of the work is making satisfactory progress, with every prospect of completion in readiness for operation next summer." Hugh Ryan and Co. had "Section 1" well in hand and reported that as much work had been completed as was possible before the removal of the land dam at the upper end of "Section 1": "The object is to reduce the quantity of material in the dam...so that after the completion of the work below water level on the adjoining section, a channel through the dam can be speedily cut...."[76] While the statistical analysis of the progress continued to include accounts of the excavation, the greater emphasis on the materials used indicates the progress of construction of the lock, piers and other works. Only 27 057 cubic yards of material were taken out of the channel; 54 252 linear feet of timber were required for cribs, stringers, binding pieces, ties and cap pieces. Other materials used included 1639 blocks, 17 749 pounds of bolts and spikes, 5943 cubic yards of stone filling, 5158 cubic feet of pine, 25 998 board feet of pine planking and 8 mooring posts.

Of course, contractual "Section 2" remained the key to the entire operation. Early in the fourth report year, on September 15, 1892, the "first stone in the construction of the lock was laid...and building was continued until 12th November, when the severity of the weather stopped operations, 7,707 cubic yards of masonry having been laid."[77] Prodigious amounts of rock and soil were still being moved from the excavation, and large quantities of materials were delivered for use in the fourth and fifth years of construction, with stone, timber, iron and cement dominating the lists. The report for the year ending June 30, 1893, was as detailed as those of previous years. A further 99 604 cubic yards of earth and rock were excavated from the prism and lock pit, 10 046 cubic yards of Portland cement were placed in the lock bottom and 28 792 cubic yards of lock wall masonry erected. As in previous years, 1892-93 saw vast quantities of material arrive on the site, including 640 536 pounds of iron bolts for the grates and culverts, 109 619 board feet of planks for the mitre sill platform, and 52 567 linear feet of timber for the prism revetment. Smaller quantities of stone, sand, cement and oak were also delivered.

Work proceeded reasonably quickly, but not without some setbacks and incidents. For example, late in June 1892, a neighbouring Indian band complained about a Hugh Ryan and Co. dynamite factory that had been established on St. Mary's Island near their houses. The dynamite was being manufactured for use on the canal. The Indians felt their lives and homes were threatened and wanted the dynamite moved farther away.[78] W.G. Thompson noted that the factory was located some 1500 feet from the Indians' dwellings, which were no nearer to the potential hazard than were the contractor's own buildings. Nonetheless, Ryan was advised to move the plant to a more suitable site. The contractor agreed, but made it clear that the Department of Railways and Canals would have to pay the cost.

In the fall of 1892 the department convinced Hugh Ryan and Co. to agree to certain revised completion dates for various components of the work on "Section 2." The expected opening date of July 1, 1894, necessitated the completion of several elements by December 1, 1893: the lock walls, wooden culverts at the bottom of the lock chamber, and the extra depth of 4 inches throughout the canal prism. Hugh Ryan and Co. also undertook the contract for preparatory work for the replacement of the railway bridge above the lock site by a swing bridge to allow the passage of the shipping using the new system.[79]

12, 13 Sault Ste. Marie Canal under construction. (SSMCC)

14 At the Sault Ste. Marie Canal *(left to right):* William Bermingham, contractor's engineer; John Ryan, contractor; W.J. Thompson, resident engineer (chief); ? Gregor, inspection board; ? Hobson, inspection board; H. Ryan, contractor; Thomas Keefer, inspection board; M.J. Haney, contractor; William Crawford, resident engineer. (Parks Canada)

At the upper entrance, "Section 3" approached completion by the close of the fourth year of construction. The beacon was finished and the dredging virtually completed (except for the portion of the channel covered by the earth dam at the lower end of "Section 3"). The entrance piers were well advanced but could not be completed until after the canal was flooded.[80] Again the statistical summary of the work accomplished at the site indicates the extent of the demands placed on the men involved. Clearly, the work was winding down. A mere 13 809 cubic yards were dredged out of the channel; 15 776 linear feet and 8614 cubic feet of timber were required; 36 745 board feet of planks, decking, "hip joists" and "girts" were used, together with 1211 blocks and five mooring posts; 6080 cubic yards of stone were dumped as fill and 11 345 pounds of iron bolts, spikes and screw bolts were used by the project.

On January 18, 1893, Allan and Fleming again asked for the return of their drawback money on the grounds that their deposit of $16 300 was "ample security for the due completion of the balance of the contract."[81] This time their bid was successful.[82] Superintending

Engineer Thompson's report for the fourth year influenced the decision. In his conclusion to the report he remarked, "there is still much to be done, but the energy displayed by the contractors, gives good grounds for believing that they will be equal to the task of completing their work by 30 June, 1894.[83]

The Fifth Report Year, 1893-94: The Lock Opens

The fifth report year ended on June 30, 1894, 1 day before the scheduled completion of the work and the formal opening of the Canadian canal and lock to navigation. This was a fateful and frantic year, but the annual report referred also to substantial accomplishments and evidence of imminent completion: "The work of construction of this canal is practically completed, excepting the river reaches, which have only been dredged out for a depth of 18 feet of water at extreme low water, whereas the lock and prism of the canal are constructed for a 20 feet navigation."[84] Water was "let into the lower level of the canal" on September 27, 1894. The guard gates were hung and on October 15 water entered the upper level and a private, if momentous, event occurred: "the steam tug *Rooth* was locked through by hand, she being the first vessel to pass through the canal, the machinery for operating the valves and the lock gates not then being in operating condition." The passage of the little *Rooth* was a signal demonstration, perhaps even a well-orchestrated one, that the completion of the works was at hand.

15 The tug *Rooth* being locked through by hand. (SSMCC)

Certainly the Department of Railways and Canals, especially its engineers who had been so closely associated with the project, were justly proud of what had been accomplished thus far. The official who penned the department's annual report commented, "I believe it to be one of the finest works of its kind on this continent, reflecting credit on the several contractors engaged in the work, and of the government in charge."

There was still much to be done before the Canadian lock could be said to be completed in all aspects of its original design and concept. After the work of dredging, blasting and excavation of the "cut" for the canal and its lock was virtually completed and after the necessary masonry and other construction work was in place, attention turned to the installation of the hydroelectric power plant and regulating machinery and equipment. The original contractors, Hugh Ryan and Co. and Allen and Fleming, were joined by several others during the last year of activity. Messrs. Beatty and Sons installed the lock pumps; Kennedy and Bros. built and installed the water-wheels of the power system, while the Canadian Machine and Engine Co. installed the machinery for operating the lock gates and valves; Miller Bros. constructed the motor houses and lock gates. An anonymous civil servant engineer referred to the lock gates as "a splendid piece of work."

On April 13, 1894, Messrs. Kennedy of Owen Sound obtained the crucial contract "for the two 45-inch new American water-wheels, to furnish power for pumping, and for the electric plant for operating the lock gates and valves...."[85] By July 1, 1894, a mere 78 days later, this work was "well advanced" but was not completed until a few weeks later. A key contract, completion of the electric light and power plant, went to the Canadian General Electric Company on May 9, 1894, to be completed by July 1, 1894 — a mere 53 days. This was a key unit in the application of electricity to the operation of the canal: electrification of the Sault Ste. Marie Ship Canal was the most technologically innovative part of the entire project. It comes as no surprise to learn that "circumstances beyond their control prevented the progress that was desired, and at date of 30th June the work was but half done." The Hamilton Bridge Company was awarded the contract for "the swing bridge to carry the line of the 'Soo' branch of the Canadian Pacific Railway over the Sault Ste. Marie Canal" on October 10, 1893. The bridge was "practically completed" by June 30, 1894.

As the lock and canal approached completion, an incident occurred that demonstrated the problems with the various demands being placed on the locale of the rapids. In January 1894 Ryan wrote to W. Crawford, the resident site engineer for the Department of Railways and Canals, informing him of a potential threat to the nearly completed canal. While the "Section 2" works had been in progress, the Sault Ste. Marie Light and Power Co. had also been constructing a power canal to the north of the canal reserve. Ryan feared that this improperly sealed canal, if allowed to be watered, would leak through the strata to damage the ship canal and he did not want to be held responsible for future damage. Because the power canal was on Crown land, the minister agreed to block the watering of this private venture if it posed any threat to the ship canal.[86] A decade or so later, after the flooding of the basements of several of the buildings on the canal reserve, Superintendent Ross must have wished that more attention had been paid to Ryan's warning.[87]

It is difficult to determine exactly when such a large project was finished; for example, the following contracts were awarded after October 15, 1894:[88]

J. and R. Miller	Feb. 1, 1895	1 pair spare lock gates
H. Flemming and Co.	Feb. 14, 1895	timber booms
Dominion Bridge Co.	March 27, 1895	moveable dam and swing bridge
Wm. Kennedy and Sons	June 10, 1895	horizontal new American turbine
Hugh Ryan and Co.	Oct. 15, 1895	six motor houses
J. and R. Miller	Dec. 10, 1894	pontoon lock gate lifter
J. and R. Miller	Unsigned at report date	office and workshop

The Final Report, 1894-95: Project Completed

By October 1, 1894, some $2 823 498 had been expended on the works at the Sault. It was anticipated that the eventual cost for the complete canal and lock system would be $4 million. Of the remaining $1 176 502, over half ($600 000) was to be directed to the further deepening of the upper and lower approaches to a minimum draught of 20 feet.[89] The remaining funds were required for the completion of the lock with all the equipment, machinery and buildings necessary for its operation and management. In short, by the close of the 1894 working season the entire project appeared to be close to completion and ready for operation.

July 1, 1894, "Dominion Day," was the projected target date and the annual report for the year ending June 20, 1894, provides a detailed statement of the work accomplished and the general state of readiness of the new system. "Section 1," the lower and eastern approach, originally scheduled for completion by April 10, 1892, was at last ready. A 150-foot section across the channel had been left as a dam at the upper end of the lower entrance to allow the completion of the work on the lock there; a few "high spots" in the channel still needed attention. The funds saved by dispensing with the beacon had been used to extend the east end of the north pier by 390 feet; the completion of the work waited on the flooding of the channel. Only some cleaning up was needed along the 5300-foot lower entrance. Otherwise "Section 1" was completed. The accompanying statistics express the work accomplished:

Dredging, deepening channel (cubic yds)	170 199
Timber for cribs (linear ft)	91 530
Binding pieces (linear ft)	6 258
Blocks under ties (no.)	4 586
Wrought-iron bolts (lbs)	60 465
Pine in superstructure (cubic ft)	14 311
Ties in superstructure (linear ft)	22 065
Blocks in superstructure (no.)	1 541
Stringers for top covering (linear ft)	3 900
3" pine planking (board measure)	54 222
Cap pieces (linear ft)	2 442
Mooring posts (no.)	16
Stone filling (cubic yds)	23 939
Pressed spikes (lbs)	2 361

Work continued on "Section 2" also. In spite of the substantial work involved in the original contract, Hugh Ryan and Co. undertook the "deepening of the prism of the canal above the lock up to...the west of Section No. 2...22 feet of water at the lowest recorded water surface above the rapids, instead of 18 feet as originally intended" and "the construction of the masonry piers and abutments for the railway swing bridge, to replace the railway trestle crossing the line of the Canal." By the deadline date of July 1, 1894, the contractors had virtually completed the lock masonry, the excavation of the prism and the construction of the supply and discharge culverts. The railway swing bridge masonry had been finished, and the side walls of the prism were nearly half completed and could be completed easily during the remainder of the construction season. Work completed to that point was reported in the usual manner:

Chopping, clearing, grubbing (acres)	30
Earth excavation in side trenches (cubic yds)	2 507
Rock excavation in side trenches (cubic yds)	74
Earth excavation in prism (cubic yds)	209 394
Rock excavation in prism (cubic yds)	126 790
Additional rock excavation in prism (cubic yds)	43 590
Quarry waste (cubic yds)	38 074
Earth excavation in lock pit (cubic yds)	42 904
Filling behind lock walls (cubic yds)	29 889
Rock excavation in lock pit (cubic yds)	150 932
Portland cement Concrete (cubic yds)	17 504
Pine timber for mitre sills (cubic ft)	4 852
Timber in culverts (cubic ft)	121 513
Drilling holes for bolts (linear ft)	4 070
Horizontal bolts (linear ft)	654
Planks in culvert floors (board measure)	333 742
Wrought iron in mitre sills (lbs)	1 568
Wrought iron in bolts in culverts (lbs)	417 820

Pressed spike in culverts (lbs)	84 133
Masonry in lock walls (cubic yds)	67 843
Excavation for movable dam (cubic yds)	1 126
Stone revetment wall of prism (cubic yds)	608
Timber revetment wall of prism (cubic yds)	36 809
Floor under revetment wall of prism (cubic yds)	95
Excavation for railway swing bridge (cubic yds)	951
Masonry for railway swing bridge (cubic yds)	2 483

As vast in scope as was Ryan's original contract for "Section 2," he nevertheless undertook other responsibilities for the project on the St. Mary's Island site. Thus, on December 19, 1893, the company was contracted "for the steel-power tube, 6'8" inside diameter, to convey water from the upper reach, to the turbines in the power-house, at the lower end of the lock, also the necessary valves, and discharge pipes from the turbines and pumps." This project was within a few days of completion by the original target date of "Dominion Day," 1894. Also, on December 19, 1893, Ryan obtained the contract for "the five pairs of lock gates."[90] This work was subcontracted to Roger Miller, who had the work in hand by July 1, 1894. On February 8, 1894, Ryan undertook "the eight valves for the supply and discharge culverts...." Bertram Engine Works Co. of Toronto had nearly completed its part of this project by deadline day. On January 26, 1894, Messrs. Beatty of Welland obtained the contract for "the two pumps for unwatering the lock...." and completed this work during the spring of 1894.

The original deadline for the upper approach to the lock, "Section 3," had been May 20, 1891. Three years later the work was still

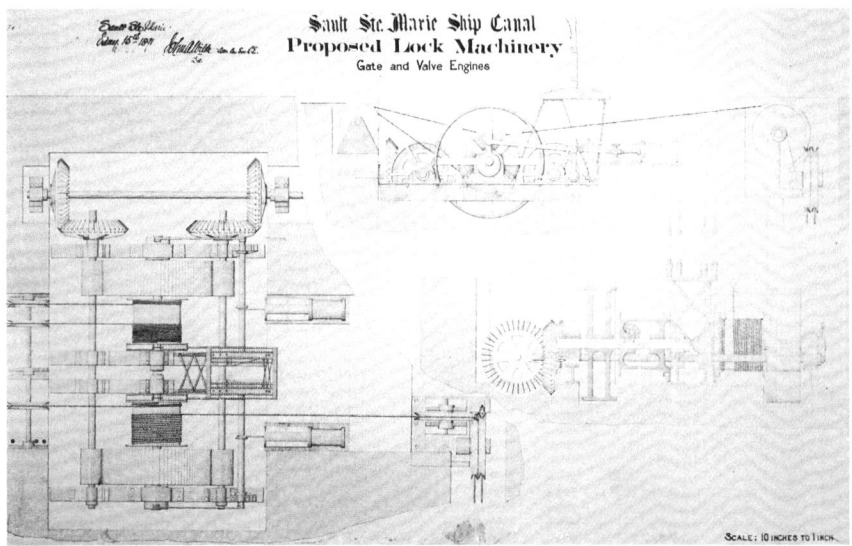

16 Proposed lock machinery — gate and valve engines. (SSMCC)

17 West side of powerhouse showing shut-off valves for conduits to turbines. (Parks Canada, SSMCC)

incomplete. The bottom had yet to be cleaned up, and 400 feet of the south pier had not yet been built because the temporary coffer dam still had to be removed. The last year had not been wasted, however, and considerable finishing work had been undertaken. The cumulative totals reveal the extent of the effort expended on "Section 3":

Dredging, deepening channel (cubic yds)	252 588
Timber for cribs (linear ft)	38 313
Timber for ties, stringers, bottoms (linear ft)	36 862
Binding pieces (linear ft)	3 574
Blocks for ties (no.)	1 968
Wrought-iron bolts (lbs)	26 254
Pine in superstructure (cubic ft)	9 755
Pine in ties (linear ft)	12 448
Blocks for ties (no.)	1 012
Stringers for top covering	2 575
3" pine plank for covering piers (board measure)	29 950
Cap pieces (linear ft)	609
Mooring posts (no.)	5

Stone filling (cubic yds)	10 025
Pressed spike (lbs)	758
Rock elm in beacon (cubic ft)	6 959
Cross ties in beacon (linear ft)	10 835
Blocks for ties (no.)	426
Pine plank (linear ft)	448
Elm plank (linear ft)	192
Stone filling (cubic yds)	2 055
5-inch sheeting (ft. b.m.)	16 972
3-inch pine plank top covering (ft. b.m.)	3 885
Hip joints and girts (ft b.m.)	2 914
Iron in bolts (lbs)	15 170
Iron screw bolts (lbs)	885
Iron straps (lbs)	5 984
Pressed spike (lbs)	269

The sixth report season ended on June 30, 1895. The main work of constructing the entrances, lock and canal prism was substantially completed by the summer of 1894, but much remained to be done, including "putting in the machinery for working the lock gates and valves, raising and hanging the gates, excavating the large dams necessary for the construction of the lock and finishing up the many details of a work of this size."[91]

Work continued on the lock itself. The main gates were ready for installation, as were the auxilliary gates and guard gates for both ends of the lock. Also, looking to the future, a contract was awarded to "Messrs. J. & R. Miller for building a spare set which will soon be completed." The two sets of guard gates were to be used in emergencies and when the lock was pumped out for maintenance. There were also some repairs to be done. During the winter of 1894-95 the prism walls were badly damaged. The lower portions were of cribwork and were underwater. When the frost left in the spring the walls were found to be bulging and required strengthening with timber braces and concrete.

The mounting of the gates and the completion of the upper and lower piers waited on the watering of the canal prism and lock. The upper and lower dams had to be removed first. As the temporary dam in the upper entrance was cut away, water entered the main works through a natural fissure in the upper dam, resulting in "much trouble and anxiety." On September 24, 1894, however, the lower reach was flooded and the lock gates hung "by means of a floating pontoon...." By October 10 the lower gates were in place and the two dams were then completely removed and the channel dredged. The upper dam was dredged easily and quickly, but the lower dam delayed the work because it was composed largely of rock. Both contractors and government inspectors were delighted when the entire work was flooded: "It is a matter for congratulations that in a work of this magnitude and built so quickly, that no leaks of the smallest consequences were found when the water was let in and the contractors deserve great praise for the material and workmanship invariably used."

After the system was flooded, work could proceed on the piers at the upper and lower entrances. Eventually the lower piers consisted of a 1260-foot north pier and a 930-foot-long south pier. At the upper entrance the north pier was 450 feet long and the south pier 1250 feet long. These four piers provided almost 4000 feet of mooring space, but even this was to prove inadequate in the early years of operation.

Despite the prodigious work accomplished at the canal and lock site, the most innovative work was accomplished as construction ended. The application of electrical power to the operation of the locking system was a technological advance, especially at the scale required. An improvement over the more usual hydraulic system used by other canals of the day, electric power promised more freedom from interruptions of operation caused by freezing and also afforded the extra benefit of on-site power for lighting the work and navigation areas.[92]

The primary function of the electrical equipment, however, was the operation of the massive lock gates, the lock valves and the pumps for unwatering the locks. The gates were 44 feet 6 inches high, 37 feet wide and each leaf weighed 87 tons. With two sets of main gates and one set of auxilliary gates, there were six leaves in all, each leaf being operated by a separate 25-h.p. motor housed in a wooden motor house. The motors were governed by "automatic switches" which allowed a simple form of remote control:

> Cords run from the switch handles to pulleys on the ceilings, and by these are conducted to the controllers and the switches are closed by the motorman pulling the cords without having to leave his position.[93]

The electric system was also applied to the filling and draining of the lock.

The water required to power the turbines in the powerhouse was brought in by a pipe 6 feet 8 inches inside diameter.[94] In the powerhouse, two 45-inch horizontal turbines (150 h.p.) drove the generator. They could also be attached to two 20-inch centrifugal pumps for emptying the lock at 32 000 gallons per minute. A smaller 13-inch horizontal turbine drove a 3-k.w. dynamo which provided the electric lighting for the entire site. The lighting system was highly advanced for 1895, consisting of "a 40 light 9-1/2 ampere Wood arc machine, which at present supplies 33 arc lamps of 2,000 candle power each spread along both sides of the canal, with reserve for additional lights when the entrance piers are finished and for range lights at both entrances." Submarine cables carried electric power under the canal to the motors and lights on the south side.

When complete, the Sault power plant and its diverse applications constituted one of the most advanced electrical systems in the world. The site became an important stop for visitors to Canada and enquiries about the operation of the system came from afar. Thus, the minister of Public Works passed on to the Department of Railways and Canals an enquiry he had received from the consul general of the Netherlands, a nation obviously interested in the application of new technology to problems of water control:

> My Government asks me to procure particulars regarding the mechanical appliances for moving the sluice doors of the lock in the Sault Ste. Marie canal, which, as I understand, are moved by means of electricity. I should deem it a great favour if you

would furnish me for that purpose with full particulars regarding this arrangement, if possible with blue prints; also with calculations of the ultimate cost of the work.[95]

18   Electric panel in the powerhouse. (SSMCC)

J.B. Spence, chief draughtsman, Department of Railways and Canals, concluded that electricity would not be much more expensive than hydraulic power over the short run, and considerably cheaper over the long run. Spence then designed the machinery and valves needed for operation of the lock gates. He also designed the powerhouse that was to accommodate the dynamos and electrical plant.

The Cost

By the end of the summer of 1895, the upper and lower channels had been dredged, well marked and were navigable. The Sault Ste. Marie ship canal was ready for business on June 30, 1895. Both channels could accommodate vessels of 17 feet draught on September 9, 1895, which was the effective date of the opening of the canal to regular shipping.[96] It was time to take the final reckoning. The chronological pattern of expenditure to this point indicates the tempo of activity at the site:[97]

Years ending June 30	$
1888	8 145.06
1889	34 018.95
1890	176 568.55
1891	325 336.33
1892	341 474.31
1893	589 801.25
1894	1 316 529.29
1895	466 151.50

The financial statement for the year ending June 30, 1896, showed that little construction was being carried on, the total for that year amounting to a relatively modest $280 711.46, of which $16 074.70 was for the support of the "staff and maintenance."[98] A final and complete reckoning was possible on November 16, 1897, which can be taken as the end of the construction period. Apart from the construction costs, Walter Shanly's award of compensation to Hugh Ryan and Co. must be included, together with an allowance for interest. Thus, the canal's total capital costs could now be assessed:

Construction to Nov. 16, 1897[99]	$3 660 459.37
Shanly arbitration award[100]	$ 211 505.00
Interest on above[101]	$ 15 844.12
Grand total	$3 887 808.49

The original budget for the project was revised to $4 million, and the Department of Railways and Canals could be proud that not only had they produced a remarkable feat of engineering and made a significant contribution to the advancement of the new technology of electrical power, but had done so $112 191.51 under budget. It was a remarkable achievement.

Completion and Opening

The steam tug *W.A. Rooth* was locked through the Canadian ship canal by hand on October 15, 1894. The first commercial vessels passed through the fully operating system on September 7, 1895. The official opening took place with due and appropriate ceremony on September 9, 1895. Canada had acquired a major new facility. In an age accustomed to mega-projects in the billion dollar range, the $4 million allocated (if not all spent) for the Sault canal must appear modest. It was a significant government expenditure for the late 19th century, however, and it stretched both the financial resources of the nation and administrative services of the bureaucracy.

If in the final analysis the bureaucrats did well, especially in making the courageous and avant-garde decision to employ electricity, the real heroes of the project were the contractors and their labour force. The contractors were largely true improvisors who faced a myriad of unexpected problems and who were expected to solve them on the spot immediately. This they did. The workers were part of this process. Their working conditions must have been horrendous but they produced excellent results. An average of 400 men worked on excavation during the peak construction years. At least another 200 men worked in the quarries while 200 more served as stone cutters and masons at the canal site. Machinists, electricians, carpenters, enginemen and dynamite experts were more specialized crafts mixed in with the general masses of hauliers and labourers. The site was dominated by horses, stone crushers, rail lines, travelling derricks, concrete mixers and an endless movement of great masses of diverse construction materials.

Scenes like this were typical of the late 19th century, which was confident concerning the application of new technology to the pursuit of "progress" and material improvements. Such mundane tasks of improving navigation became elevated into national projects for the betterment of the society and nation at large. But there were a few local losers. Squatters had lived on St. Mary's Island for several years and these poor, deprived people were driven off. The pathetic nature of this process is made clear by a Mrs. Crowley's appeal to the Hon. A.G. Blair, Laurier's minister of Railways and Canals:

Hon Sir
> I write for to let you know that there are ten Families of us living the Canal grounds Names as following Edward Hosmers Healeys Pelans Bennett Patterson Jacobson two Goolie [i.e. Icelandic] families and myself Mrs. T.H. Crowley

Hon Sir
> Boyd Superintendent and Lawyer McKay has served me with paper to move of but I am not able for to move my House or buy a lot place it upon but if you will grant me some assistance I will move my house as soon as possable my Husband has gone west to work on a rail way and I have a large family and no money to move at present I trust that your department will grant me some recompence as soon as possible.[102]

Above all, the ship canal was a Canadian achievement. Some of the arguments in favour of it, especially the one of national security, may have been specious; the economic argument was overstated; and the basic canal design was almost certainly faulty. Nonetheless, the engineering was brilliant and the various components proved that Canadian, especially Ontarian, industry and design had come of age. The study of comparative canal technology is yet in its infancy; nonetheless, some points are clear. The Sault Ste. Marie project was the first Canadian canal to be electrified. Spectacular use was made of the technology half a century later when the entire St. Lawrence Seaway system was also electrified. James B. Spence and his associates rank as major figures in the history of Canadian technological innovation.

IV OPERATING THE CANAL

On September 7, 1895, Captain Peter Campbell, commodore of the Great Northern Transit Line, took the new Canadian passenger steamer *Majestic* with about 700 passengers aboard through the recently completed Sault Ste. Marie canal. Although this was the first ship to actually negotiate the new system, the official opening took place 2 days later on September 9 at 7:00 AM when the canal was "formally opened for public business."[1] Although J. Boyd, the first canal superintendent, thought it "appropriate" that the first ship through was the Canadian *Majestic,* the passage of the two American steamers *Uganda* and *City of London* at the official opening is both ironic and symbolic. It is ironic because a system that was advocated for the better development of the Canadian transport system and the guaranteed passage of Canadian vessels and commerce was thus first used by American steamers. It was symbolic because these two ships represented the vanguard of the American fleets which were to dominate the traffic through the Canadian Sault in the future.

Despite the nationalistic sentiments prompting their construction, both the American and Canadian sets of locks came to be integrated into one continental system. Indeed, if the operation of the Canadian Sault is to be fully appreciated, it must be considered in the context of the capacity and operation of the whole system which included both the Canadian and American canals.

Operating the New System

Apart from the length, width and depth of the locks and approach channels, any evaluation of the effectiveness of the system must also consider the efficiency with which it operated. Poor handling, equipment failure, accidents and slow progress through the locks could affect the pattern and degree of use of the various locks. The first year of operation of any canal was always potentially problematic. The training of a new operating crew was made more difficult by the novelty of the electrical system that operated the lock and powered the establishment on the site. The Canadian canal was innovative in its pioneer application of electrical energy to the operation of the locks while the decision to construct a long narrow system rather than the wider prism which would have allowed double lockages was also under scrutiny. The first few seasons tested both these decisions.

In July 1896 Boyd employed some of the operating staff in "cleaning up the machinery and fixing up about the power house." After a month of this familiarization and preparation, Boyd reported that "I put these men to work operating the machinery which is all done by electricity and as they were all green hands it required a good deal of instruction to teach them their duties. It was time well spent as is shown by no accidents of any amount having occurred from the operating of the machinery." The first day's operation saw the passage "without mishap" of some 41 vessels with a combined tonnage of 44 469 tons in 9 hours. Reporting on the first operation of the electrical plant, one professional journal commented, "The lock, lock gates and power house and all the valve and gate machinery were designed by Mr. James B. Spence, of Ottawa...and the uninterrupted smoothness with which the entire work has operated since the opening of the new lock, indicates the thoroughness with which every detail has been worked out."[2]

At the close of the first year of operation, the only major criticism was that the dynamo that generated the power for lighting the buildings "has proved to be too small for the work required of it, especially during the long nights in the autumn when the strain on it is very heavy."[3] The locks were operated 24 hours a day during the navigation season and the 33 arc lamps and 7.3 miles of wire, "including 4,250 feet of submarine cable," placed too great a strain on the original equipment. The latter was thought to be "unadapted to work in this country where we have so much wind, rain and darkness" and it was essential that the dynamo not break down as "with the strong currents which cross both of our channels it would be very dangerous to vessels in the channels if the range lights were to go out." A replacement was recommended and installed. In 1911 another electrical problem was reported in that "the difficulty of controlling the machinery in the power house under the greatly varying load...caused the power to fail at times when the maximum load was required."[4] A new water-wheel and modifications of the intake and discharge pipes were expected to resolve this problem.

Apart from these minor teething problems, the power plant worked well. Indeed, it was not until 1942 that a major modification of the original system was effected:
> Commencing at the close of navigation, a complete change-over of the electrical system for operating the lock valves and gates and for lighting the canal was undertaken in preparation for the 1943 traffic. Generation of direct current at the canal power house has been abandoned and alternating current is being obtained from the Great Lakes Power Company on a contract basis. New conduit, wiring and motors have been installed throughout.[5]

The proposed alterations included six 10-h.p. "wound rotor motors...one for each gate...which allows any gate to be individually operated or any pair of gates to be operated together from the south side of the lock," four "squirrel cage motors" for valve operation; and a new lighting system that included "the provision of red and green signal lights for traffic control through the lock."[6] The following year's report noted that "A completely modern electrical system of operation and lighting was installed on this canal early in the season, replacing the direct current system which had operated since the canal was first opened in 1894."

From the beginning, the new lock was also thought to be functioning well and fulfilling the expectations of its proponents:
> By the opening of this canal the congested state of the traffic on the St. Mary's river was relieved, previously to this vessels had

19   The canal in operation. (SSMCC)

been obliged in some cases to wait from twelve to thirty-six hours for their turn to lock through the American canal. Such protracted delays are now unheard of and the delay of an hour is now considered to be a great hardship.[7]

The improved traffic flow was thought to reflect the superior design of the system:

The long narrow lock has proved to be a great success, the work of passing through it is done with much greater dispatch than could be the case with a wider one, and the correctness of the views of those who changed the plan from the old to the new form of lock is fully demonstrated. There is no time lost in placing vessels alongside of each other as would be the case of a shorter and wider lock, and the use of a tug for that purpose is reduced to a minimum, and no damage can possibly arise to vessels in surging across the lock and hitting the opposite side. In this long and narrow lock a steam barge enters with her consorts without the assistance of a tug as was exemplified in one case this season where a barge brought in four schooners without the assistance from a tug.[8]

The initial design had called for a theoretical lockage of one upper laker and two smaller vessels. The average lockage for the first year of operation was 1.68 vessels per lockage. The other consideration was the innovative power plant:

> That the operation of the lock by electricity is vastly superior to that by hydraulics, is proved by the quicker despatch in locking vessels, and the ease in operating our lock as compared with the locks on the American side, both of which are operated by hydraulic power, our lock having been in operation without interruption or without causing delay to vessels from the date of its having been opened for traffic in 1895 up to the present time [1897].[9]

This initial first flush of enthusiasm for the operation continued to 1898, it being reported that

> the adoption of the long narrow form of lock, and its operation by electric power has been proved by the quickness of despatch in locking vessels through to be the best form of lock adapted to the traffic as compared with the wider form of lock on the American side which is operated by hydraulic power. The average time of making lockage, including all delays to vessels in this lock, amounting to only fourteen minutes and fourteen

20 Whalebacks passing through Sault Ste. Marie Canal. (SSMCC)

21 The canal in operation. (SSMCC)

seconds, whilst that of the American lock averaged thirty-six minutes and thirty-one seconds.[10]

Naturally, the speed of lockage was not so remarkable during the first few months of operation although Boyd reported that "The operating staff as a whole were new to the work and so far they have done as well as could be expected considering their experience at such work."[11] The "green hands" of 1895 returned an average lockage time of 18.26 minutes for the 698 lockages of the 1193 vessels and 748 371 tons of that year; the record performance was attained in 1897 with an average of 13.97 minutes for the 2976 lockages of 4376 vessels of 3.8 million tons.[12] Thereafter, at least this measure of the efficiency of the operation decreased, the average time of lockage gradually increasing year by year to 20.60 minutes in 1908.

The principal indicator of the efficiency of the canal and its locks, therefore, was overall average time per lockage. Ross, the second superintendent, provided detailed statistics of the time it took ships to proceed through the locks:

Downbound vessels

From stop post to lock	7 min.
Moving into lock	8 min.
Lowering vessel	11 min.

Moving out of lock	4 min.
Clearing lower stop post	4 min.
Total time	34 min.

Upbound vessels

From stop post to lock	7 min.
Moving into lock	10 min.
Raising vessel	10 min.
Moving out of lock	7 min.
Clearing upper stop post	5 min.
Total time	39 min.

These figures demonstrate clearly that the lockage itself occupied only a third of the total time required to negotiate the lock, most of the time being spent handling the vessel. The lockmaster had to be patient because any interference to attempt a speedier passage could result in the "responsibility for any accident that might occur [falling] upon the canal authorities rather than on the master of the vessel."[13]

The length of the season of navigation also affected the efficiency of the operation. From 1895 to 1916 the shortest navigation season was in 1896 when the Canadian Sault was open for only 218 days whereas the longest season was the 264-day navigation period of 1902. The general rule of thumb was an 8-month navigation from the end of April to early December.

Within that period, the lock activity varied throughout the year. The lockmaster's day journal for 1913-14 indicates the seasonality of the traffic.[14] Navigation opened on April 25 with the passage of the first of about 36 vessels, the *J.G. Munro,* These must have been waiting for the system to open as the number then decreases, although by mid-May the traffic has increased to daily lockages of 30-50 vessels per day. The peak day is July 26 when 77 vessels were locked through, and about 329 in total for July 22 to July 29. Not until December does the flow slacken and the season closes on December 13 with the passage of the *B. Lyman.* The navigation lights in the approaches and floodlights at the locks allowed the canal to operate around the clock. The lock books indicate the daily plan for the peak day of July 26, 1913.

The Establishment at the Sault

A typical year in the 1905-36 period required about 50 persons to perform the various tasks essential to the continued operation of the power and transport facilities. The canal site was a busy location during the navigation season, which saw the constant arrival of vessels to be locked through. The winter was no less busy although the routine changed. The locks were drained and the equipment checked, overhauled and, when necessary, replaced. A detailed report outlines the specific tasks likely to be performed:

> During the winter...Extensive repairs were made to the lower main gates, but at the very best they can only be called temporarily repaired, and if they last out the balance of the season they will be doing well. A new pair must be built this winter and be ready for the opening of navigation. All the machinery has been thoroughly overhauled and necessary repairs were made. New valve rods were put in as the old ones were found to be too light for the work required of them. One of the pump shafts broke and upon examination it was decided they were too small, so new and larger ones were put in both pumps and new brass collars were put on them so as to do away with rusting in the the bearings and so casing trouble as was in the present case....The inside of all the buildings have been painted and next year all the outside work will require to be done.[15]

The supervision of the establishment fell to the resident superintendent, a position J. Boyd held from the opening of the canal in 1895 until 1906. In 1907 F.B. Fripp was the acting superintendent, being replaced in 1908 by J.W. LeBreton Ross whose new title was superintending engineer. Other senior staff members included the "engineer in charge," responsible for the construction and repair activities required to ensure the satisfactory operation of the locks and the uninterrupted navigation of the approaches to the locks. He was assisted in this task by the dually specialized "diver-day oiler" who was based on site at the canal. The electrical system was overseen by the canal electrician and his staff of 4 assistant electricians, oilers, machinists and motor men. The operation of the locks fell to crews of 8 to 20 linesmen supervised by the lockmaster and his foremen. In 1921 the linesmen received the new title of lockmen. The outside duties associated with manning the locks also included the crew of a patrol boat/tug to assist in marshalling the traffic into the lock and the approaches, and a 2-man lookout station which had been established at Point aux Pins. The report for the 1913-14 season praised both the station and the patrol service as "Very few vessels came to the lock out of their turn, and the congestion of traffic which usually occurred twice a week was handled without difficulty." Finally, about 20 people were employed as "scrubbers," "lamp trimmers," "hoist runners," labourers, handymen, painters, carpenters, blacksmiths and gardeners.[16] The latter became a permanent feature of the establishment after the appointment of three gardeners in 1921 and two every year thereafter.

The superintendent's "bible" for the operation of his canal and lock was the comprehensive set of "Rules and Regulations for the guidance and observation of those operating and using the Canals of the Dominion of Canada, to take effect on the 1st May, 1895."[17] Specifications and requirements regulating the passage of vessels and the conduct of their skippers and crews were presented in minute detail, together with the fines to be levied for non-compliance. Understandably, attention was directed to matters of "clearance," the provision of "whistle, bell or horn" for signalling the lock, "lights" at bow and stern, "screens" on chimneys to protect against fire, and other matters relating to the safe operation of the system. The draft of vessels was an important consideration: "Every vessel or boat navigating any of the canals shall be correctly and distinctly marked and gauged in feet and inches at the bow, amidships and stern, showing the exact draft of water drawn by each portion of the vessel or boat, neither of which will be allowed to enter ... the Sault Ste. Marie Canal [drawing more than] seventeen feet." Priorities were assigned to vessels moving through the locks. "First class" vessels were those steamers employed in "the carriage of passengers," "second class" comprised "all other vessels, of what kind soever they may be," and "Mail Steamers" which were said somewhat whimsically to "always have priority of passage over all other vessels whatsoever." After the traffic was regulated, attention was directed to the staff who were warned that "No lock-tender or other officer on the canals shall keep, or in any way be interested in any inn, tavern or grocery, nor shall he sell...any articles or property whatsoever, to any person navigating or travelling on the canals." A special circular was appended to the regulations for the benefit of the lockmaster, who regulated the water levels and controlled the vessels, ensuring that "all vessels while waiting, take and keep a safe and proper position with all needed lines out." He was required to "direct every lockage personally, paying particular attention to the opening and closing of gates and valves, and rate of speed of vessels, both entering and leaving the lock....To exercise the greatest care in locking, making sure that a vessel is securely tied up in the lock, before allowing the men to begin closing the gates...." The description of his duties as an overseer of his linesmen gives some insight into the work conditions of the day: "To see that the men do not leave work without permission. The Lockmaster has no authority to grant this permission. (There will be no holidays). If through sickness, or other unavoidable cause, a man is off for a day, he cannot be paid for that day." Finally, "The Lockmaster.... will be held personally responsible for the full, careful, and proper performance of all duties, on the lock...."

Although the locks were complete and the rules for operating and managing the lock and canal were available, not all of the physical establishment at the lock site was ready for occupation by the time navigation opened in 1895. Apart from the lock system, the canal site was unfinished. Plans called for an administration building located conveniently to oversee the canal operation. Accordingly, tenders for construction were put out in August 1895, the specifications called for a two-storey structure built of the red sandstone through which the canal had been excavated with decorative detail and trim around doors and windows in limestone. It was not until May 1896 that Boyd could occupy the "new offices," commenting that they "greatly facilitated the work of the clerical staff." In August 1897 Boyd reported "The superintendent's residence has been completed and is now occupied, and the sewer and water pipes from it are in course of being laid."[18] This building symbolizes the rank and prestige of the superintendent's position; his function was considered important and his residence befit that function. Moreover, Superintendent Ross in particular asserted that prominence and prestige in the local community, and his home at the canal site became part of the social round of the elite of Sault Ste. Marie. Indeed, the administrative building still stands as a dominant and impressive element of the canal site.

Other finishing work included painting the motor houses, plastering the interior walls of the powerhouse, and painting the woodwork. Boyd was not unaware of the needs of his men and, for several years, one of his particular concerns was that "A small frame building should be erected for the use of the motormen and linesmen when not actually operating the lock, as the room now used by them in the power house is not of sufficient size to accommodate the number of men using it."[19] Eventually a small sandstone shelter met this need. Other structures added to the site in ensuing years included a new shed "close to the shops" in 1912 and a large lumber shed at "the east end of the grounds" in 1914. During the first two decades of operation, another major modification of the site was the repeated extension of the piers at both ends of the approach to the lock. In 1919 Ross reported the following physical plant on the site:

Item	No.	Comments
Locks	1	Electrically operated
Bridges	1	Owned and operated by the C.P.R.
Emergency dams	1	Swing dam operated by hand, wickets lowered by steam
Light houses	4	Range lights at entrances powered by canal plant
Buoys	21	9 gas and 12 spar buoys
Plant maintained		Derrick scow, tug, gasoline launch, flat scow for diver
Electrical power station	1	
Buildings		1 office, 2 residences, 3 watch houses, 6 motor houses, 1 shelter, 1 shop, 2 storehouses, 2 gauge houses
Highways		3800 linear ft

A perennial concern for Boyd and, later, Ross was the state of the grounds. Initially, their attention was directed to the considerable mess of rubble and disturbed ground the contractors left; later they maintained their pressure for allocations of funds to continue the development of attractive gardens, lawns and shrubbery at the site. The very first annual report commented that "The canal grounds require to be trimmed down and levelled off, which service will require a considerable expenditure of money so as to give them a neat and tidy appearance. Around the offices a small portion has been levelled and grass seed sown.[20] Grading and levelling were continued in the following year. Successive annual reports recorded that the planting of trees and more levelling continued over the years and in 1902 Boyd argued that "A small sum set aside each year for this purpose would soon accomplish the desired end, and give our grounds a very much improved look." Other improvements requested included the replacement of the wooden walks along the locks. In May 1907 the annual report noted that cement walkways had been installed on both sides of the locks in place of the original plank walks. Indeed, after a decade of occupation of the site Boyd reported:

> The small harbour crew kept on hand at general work have been engaged in levelling up the ground when not otherwise employed, and their work is now beginning to show and a small annual grant should be made for that purpose so that the grounds could be got into shape so as to be fitted up in comparison with those on the American canal, which are a delight to the eye and not a mass of rough rock and grounds as ours now are.[21]

Apparently, the two canal systems were to compete in aesthetics as well as economics, but as late as 1908 Ross feared it would take another season to level the "spoil banks." He and Boyd had been striving to make the locks a "place of beauty." Certainly, although not formally a park, they were becoming a place of public resort and it was recommended in 1908 that certain additional developments were required because of this:

> It would seem advisable to place a pavement along the north side of the lock, outside the snubbing posts to accommodate the public, as the crowds of people who visit the lock at times interfere with the men operating the lock and also run considerable risk of accident, as the lines from boats at present cross the pavement.[22]

Just as the rapids had attracted the attention of the local population and visitors alike as a place of splendour and beauty, so the canal site attracted the public because of its interesting activities and the attractive grounds. Government had allocated the funds for the concept of the canal, the contractors had presented Canada with the finished product of power system and lock system, but it was Boyd and Ross's initiative that gave the community the park-like setting so much needed in an industrializing Sault Ste. Marie in the early 20th century.

Finally, there was one noteworthy activity which although not administered by the Department of Railways and Canals was nonetheless integrated into the commercial system at the Sault. Postal, telegraph and telephone communications were added to the facilities there at an early date. In 1898 the C.P. Telegraph Department requested permission from Collingwood Schreiber, deputy minister and chief engineer of the Department of Railways and Canals, to open a telegraph office at the canal.[23] In May 1909 a post office was opened at the canal.[24] Before that date the Sault ship canal post office had been only a "summer office," open only during the navigation season. The following year it was elevated to the rank of "Accounting Office" and given the added responsibility of issuing and cashing money orders and postal notes. From

1910 until 1933, Superintending Engineer Ross added the duties of postmaster to his other responsibilities in return for an annual salary of $100. Apart from being a convenience for the better administration and management of the canal, these various communication and financial services were useful to the shipping interests and the post office operation became part of the daily activity at the lock site.

Collisions, Groundings and Blockages

Despite all the efforts of the successive superintendents to ensure an efficient operation at their canal, accidents did occur and the traffic was frequently interrupted. Two major types occurred in the half century or so of operation before 1945: collisions and groundings while approaching the lock, and collisions while being locked through the system. Both were equally serious as they could block the flow of traffic, causing considerable congestion at the locks remaining open. In addressing this question, Superintendent Ross recognized the primacy of "the safety of the lock and of the vessels using it" but was also concerned that "at a lock where the traffic frequently becomes congested, the question of speed of operation should receive some consideration consistent, however, with safety."[25] He went on to estimate that in the 1912 season alone, 3216 vessels were delayed for about 10 345 hours in total; assuming a loss of earnings at the rate of $21 per hour, this amounted to a total loss of $217 245 for the season. Hence his concern. Congestion was caused by fog, and a regular weekly congestion on Sundays and Thursdays because "vessels are not allowed to load on Sundays, so that they become bunched together."

Then there were the accidents; congestion caused "by one of the canals being put out of operation by an accident" was also common. When the American Poe Lock was rendered "hors de combat" following an accident in November 1909, the Canadian lock felt the pressure, being in continuous operation for 264 hours to pass the 460 vessels having a total tonnage of 1.37 million tons. Even with these prodigious efforts, there was at one point a blockade of about 87 vessels above the locks, each vessel experiencing a delay of between 60 and 100 hours, the total financial loss amounting to $250 000. Another interruption in the operation of the Poe Lock the following spring resulted in a back-up of 140 vessels at anchor, and the Canadian system worked continuously for 559 hours, passing through 1148 vessels and 2.88 million registered tonnage.

Preventing accidents was difficult. Apart from the many possibilities for human error, the system itself had not overcome all of the environmental hazards to navigation. Thus the upper, or western, approach was rendered hazardous by a strong current which was particulary troublesome to "vessels with tows" as demonstrated by the schooner "Nelson" which struck the end of the pier there and "sunk immediately."[26] Negotiating this current was compounded by a pier located in the middle of the channel to carry the C.P. swing bridge across the canal. Complaining that "some very close calls have already occurred, and it has now become a cause of great complaint of vessel men using the canal," Boyd recommended "the removal of this pier and the erection of a bridge swinging clear across the canal." The report for 1898 restated the case, echoing the "remarks of condemnation from vessel captains as formerly in other seasons," and calling for the removal of "what is considered by all vessel men a serious obstruction to the operation of the canal." Boyd's constant pressure could not be ignored and by 1899 he reported the construction of a new swing bridge across the canal by the Dominion Bridge Company of Montreal, and the removal of the offending bridge-pier from the channel.

After the C.P. pier was removed, attention was directed to protecting vessels from the current. Engineer Fripp recommended an 800-foot extension of the "South Pier, Upper Entrance" which would "act as a protection in keeping vessels from being driven on the bank."[27] Such was the power of the current there that despite the engineer's efforts, in 1907, the year following the improvements, the superintendent reported that three steamers, the *Harvey Coulby, D.M. Clemson* and *Hoover & Mason,* collided with the new pier, moving the end of it some 2-1/2 feet out of line.[28] Despite the improvements, the problem continued and apparently was compounded by other developments. During the 1918 navigation season, no less than 11 vessels were grounded on various occasions at the western approach, because of the "currents round the end of the pier, due to the draught of water into the power canal of the Great Lakes Power Company."[29] Again, in 1925, navigation was interrupted by several hours after "The steamer *Northwind* entering the canal from the west was drawn over by the current of the power channel and struck the corner of the north pier" before going aground.[30]

These incidents seldom interrupted the operation of the transportation system for long. Grounded vessels were towed free, damaged piers were repaired at the expense of the vessel owners and damaged ships were repaired at Sault Ste. Marie or other yards nearby. Accidents affecting the operation of the locks were more serious. This was recognized from the start and the emergency swing-dam to the west and upstream of the lock was designed to protect the system if an accident affected the lock gates and the regulation of water into the locks. The principle of operation of the swing-dam was as simple as it was important:

> The dam consists of a swing bridge structure, from which a number of frames are suspended. These frames are hinged at one end, to the bridge floor truss, the other end, when lowered, bearing against a sill at the bottom of the Canal prism. The frames, when in position, act as guides for wickets which cut off the flow of water.[31]

Boyd was as assiduous with this as he was with all matters connected with the operation of the canal and his report for 1901 records that "the swing dam was operated, or rather a part of the wickets were let down and the men instructed in its operation."[32] Efficient operation included being prepared to respond to accidents in the system. The lock gates were particularly vulnerable, as were the mitre sill and the gate cables, the latter susceptible to being cut by vessels drawing too much water passing into or out of the lock. The most dangerous accidents, therefore, were those that occurred near the lock. Some were bizarre:

> On October 17, at 4:30 a.m., while the steamer *Emperor,* 525 feet long, was locking down, the lockmaster signalled to turn on the water at the upper gates for the purpose of flooding the vessel out of the lock, before the captain was on board, mistaking another man for the captain. The vessel started ahead with the flood before her lines were let go, breaking her lines, breaking the forward chock and carrying away about fifty feet of her railing. The vessel might have been stopped without further damage, had she not dropped her anchor in the lock, with the result that the anchor went through the bottom of the vessel, and caused her to settle on the bottom right after she had passed out of the lock....After lightering [sic] a portion of her cargo the *Emperor* proceeded on her way to Port Colborne.[33]

Given the frequency of accidents, it is surprising that the canal operation was not interrupted more often and for longer periods. Not only the *Emperor* had trouble in the 1911 season: in May of that year, the steamer *A.Y. Townsend's* paddle wheel created sufficient suction on entering the lock to drag away the north lower guard gate from the wall, together with the bridge there; in June the steamer *Hamonic* collided with the lock wall, breaking the coping; in August the steamer *Isaac L. Ellwood* was struck by the steamer *Kaministiquia;* in September the steamer *Newona* collided with the lower south wooden pier; in October a week after the *Emperor* incident, the steamer *Caribou* "broke all four blades off her wheel on an old gate which had been standing against the end of the lower north pier, and which had fallen over." Despite this litany of incidents, the superintendent reported for that season that "There were no accidents of such a nature as to seriously interfere with navigation."

The Canadian Sault had not always been so fortunate, however, and 1909 saw a particularly dramatic and destructive accident. Providentially, or rather because of the efficiency of the system at the Sault, the previous winter's chores had included the refurbishing of a piece of equipment that was to prove to be essential the next summer: "The movable dam was scraped and partly painted. This work will be completed this spring."[34] Despite the concerns about the approach from the western end, the challenge came from the east at the lower lock gates. A stacatto barrage of telegrams from Ross to the ministry at Ottawa highlights the sequence of events:

> June 9, 1909
> Steamer Perry G. Walker knocked off lower gate from lower side while the Assiniboia was in the lock & The Crescent City was lying above. Both boats passed through with the current, the upper gates were carried away along with the current as well. I am securing the auxilliary & guard gates & will close the swing dam immediately.

> June 10, 1909
> All vessels collided in accident and were seriously damaged but

no lives lost. The lower entrance pier was also damaged. The swing dam is closed and wickets lowered. 7 wickets are jammed — we are endeavouring to force them down with jacks and stop the leaks.

June 10, 1909
Need Mr. Miller or Hugh Ross here by tomorrow night as the work of stopping the flow of water is difficult & progressing slowly.

June 11, 1909
All wickets down except broken ones. Am attempting to place bulkhead across opening.

June 12, 1909
Still working to force bulkhead down with jack. 4 more feet to go.

June 13, 1909
Both pairs of guard gates are safe, endeavouring to stop leaks in dam. When this is accomplished will close guard gates to pump out lock & examine sills.

June 14, 1909
Lock on water floor slightly damaged on upper end, timber on upper breast wall carried away. Culvert filled with debris.

June 15, 1909
Repairs to floors will be completed by noon today. Repairs to upper wall & cleaning out culvert should be finished by Wednesday night.[35]

The details of the story follow.[36]

At 1:45 PM on June 9, the scene at the Canadian lock was one of vessels arriving and departing, and linesmen and ships' crew going about their usual routine duties locking through vessels. The C.P.'s *Assiniboia* was already in the lock while the Pittsburgh Steel Co.'s *Crescent City* was slowly under way to enter, some 150 feet of her 426-foot length having passed the upper gate. Below the closed gates, *Dredge No. 10* was lying by the end of the north pier, waiting to be locked up when the two vessels in the lock eventually passed out of the lock. At that point, the *Perry Walker* entered the lower approach to the lock at a "high rate of speed."

There must have been some breakdown in communications between the bridge and the engine room on the *Perry Walker* because she continued her rapid approach to the closed lock gates. Seeing this, the captain sent a runner down to the engine room, calling out "which way is your engine working?" The wheel was seen to reverse but it was too late to stop the vessel's progress which was later reported to have been about 5 or 6 miles per hour when she was only 200 feet from the lock. A line was thrown out to the shore but this was a last futile effort and the

Perry Walker struck the south lower gate, forcing it back, and causing the north lower gate to fall over.

The water already in the lock rushed out, throwing the *Perry Walker* back into and across the channel. The torrential flow bore the already mobile *Assiniboia* out of the lock, crashing into the lower gates on her way through. Her attempt to reduce her velocity by dropping her anchor failed and she rushed on, striking the *Perry Walker* a glancing blow as she passed her. The *Crescent City* was in trouble too, her bows dropping as the water drained out of the lock and over her deck forward, and then "her stern went down and bow up, and water washed over her afterdeck."[37] Lines were thrown to the shore and her engines were reversed. The full force of the now uncontrolled waters of the St. Mary's River were rushing through the lock and she was flushed out at the stupendous rate of about 40 miles an hour. The *Crescent City* must have moved faster than any other laker had ever moved, but she could have done without that questionable distinction. Tearing away the timber work along the breast wall and crashing her bottom on the lock sill as she passed through the lock portal, the *Crescent City* overtook the *Assiniboia*, striking her in the stern above the water line.

The scene from the lock side must have been spectacular. Two giant ships racing through locks where heretofore vessels had inched their way along, gingerly negotiating their entry and departure under the watchful eye of the lockmaster. The view from the decks of the two vessels involved must have been even more impressive. One account of the "Thrilling Accident" describes the scene on the *Assiniboia:*

> Pandemonium reigned on the *Assiniboia* for a time after the accident. Several people fainted and others were so overcome that stimulants had to be administered. One passenger attempted to jump overboard into the seething waters, and was only dissuaded from doing so by the mate who, when other means failed, floored the man. Another passenger threw his trunk overboard. The officials of the *Assiniboia* feel that their escape was almost miraculous. In an interview, one of them stated that, had the *Crescent City* struck the *Assiniboia* in the centre below the mark the boat would almost certainly have been cut in two.[38]

Apparently, not all of the passengers were disconcerted at their unexpected "shooting-the chutes" at the lock, as "Some of her [*Assiniboia*] passengers apparently supposed this method of locking was the usual one, for they are reported to have continued to snap their cameras as they were shot along the canal."[39]

All three crippled vessels limped to the American side where divers examined them. The *Assiniboia* moved out later that day on route to Owen Sound. Three days later the *Perry Walker* continued on her way up to Lake Superior, but by way of the American canal. Even the *Crescent City*, which had sunk at the wharf after being holed by the *Assiniboia's* anchor, was raised after a few days and proceeded on her way to Cleveland, Ohio. These vessels left behind a scene of destruction and an engineering problem of major proportions.

22, 23 Sault canal accident, 1909. (SSMCC)

The first task was to control the flow of water through the lock. The lower gates had been destroyed by the vessels passing through the lower portal. The upper operating gates were open at the time of the accident but they too came under pressure from the flood of water rushing through the lock: "They were trembling all the time and an attempt was made to secure them to the walls, but before this could be done they were sucked out and carried away, breaking the gate anchorages and in one case splintering the oak heel post."[40] Both pairs of guard gates and the auxilliary gates were secure in their recesses and were not damaged. The force and velocity of the water were colossal. There was a fall of 3-4 feet in the upper canal, another of about 12 feet at the upper mitre wall of the lock, and yet another of 4 feet at the lower end of the lock. The surface velocity in the lock was about 15 feet per second, dropping to 7 feet per second in the canal.

Once the guard gates were secured, attention was turned to the movable dam, 1000 feet upstream of the lock. Here several problems had to be overcome. First, the hand gear for turning the bridge broke and a team of horses was hitched up to haul it into position. Then there was the problem of the wickets. By 3:30 PM the bridge was in place across the canal; the first of the 23 wicket frames or girders were lowered into place; the wooden wickets were then slid into position, some of them by a 100-ton jack hammer; one section was warped beyond use, so a bulkhead of lumber, planks and bales of straw was used to close the last section. By 3:10 PM on June 13 the auxilliary gates were closed and the work of pumping the locks dry began; by June 15 repair work could begin.

To the professional, "Aside from the wrecked gates the damage was comparatively slight."[41] To the uninitiated, it must have appeared extensive: the timberwork of the upper breast wall had been swept away; both thicknesses of planking were torn out of the lock bottom; 36 out of 40 culvert gratings were gone; cables for the electrical lighting system had to be replaced. By June 17 they had completed the lock repairs, let water into the lock, and began raising the dam. They hoped to proceed quickly enough dredging the channel to recommence navigation by the evening of June 18.

The aftermath of the accident was significant. First, there was some justification for pride and self-satisfaction. The prescience in building the swing-dam and the general efficiency of its operation were appreciated. Second, the experience justified strengthening the auxilliary gates and guard gates and the provision of spare main gates. Finally, Ross introduced new rules regulating the movement of boats approaching the canal; two extra linemen were stationed one at each end of the piers at the western and eastern approaches to the lock. Moreover, all vessels were now required to "come to a stop at the entrance piers, unless the gates are open for them, and to put linemen ashore," and only then were they "permitted to enter the lock with their lines ready to check the movement of the vessel if necessary."[42] There were other accidents but none as serious as that of 1909. The year even closed with a tragedy: "On December 27, while repairs to the lock were in progress, one man fell from the top of the wall down to the bottom of the lock and was instantly killed, while another man who was on the lock

24 Emergency swing-dam in place. (SSMCC)

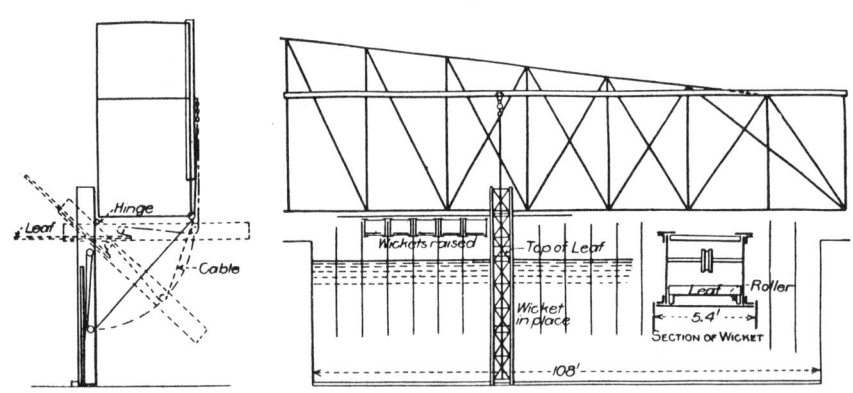

25 Plan of the emergency swing-dam. (SSMCC)

bottom had his arm broken by a falling timber."[43] Superintendent Ross must have been pleased to finish his 1909 report and put that calamitous year behind him.

Canal Traffic

Locks and Channels

The effectiveness of the Sault canal can only be evaluated by an analysis of its primary function, the handling of traffic. Although several factors have influenced the volume, direction and constitution of flows through the canal, the efficiency of any canal system is determined largely by its ability to accommodate the dimensions of its carriers. Therefore it is essential to consider the length, breadth and depth of the locks, and also the dimensions of the approach channels.

By the time the Canadian Sault canal was operational, the Americans had already completed two canals and another was in progress. The first, the St. Mary's Falls Ship Canal, was constructed by the State of Michigan and opened for navigation in 1855; it consisted of two locks 350 feet long, 70 feet wide and 9 feet deep, with a 5400-foot-long and 100-foot-wide approach channel. In 1881 the Weitzel Lock was added, which was 515 feet by 80 feet by 16 feet; another development, the Poe Lock, was completed in 1896 on the site of the original state locks and it was 800 feet long, 100 feet wide and 20 feet deep, continuing the trend to a more capacious system. The Canadian system was longer and deeper, if not wider, being 900 feet long, 60 feet wide and 22 feet on the mitre sills. The few extra inches of depth in the Canadian lock, together with a deeper approach channel and a more speedy and efficient single locking system, gave an advantage to the Canadian lock. This advantage lasted from 1895 until 1914 when the Americans added to their canal capacity at the Sault.

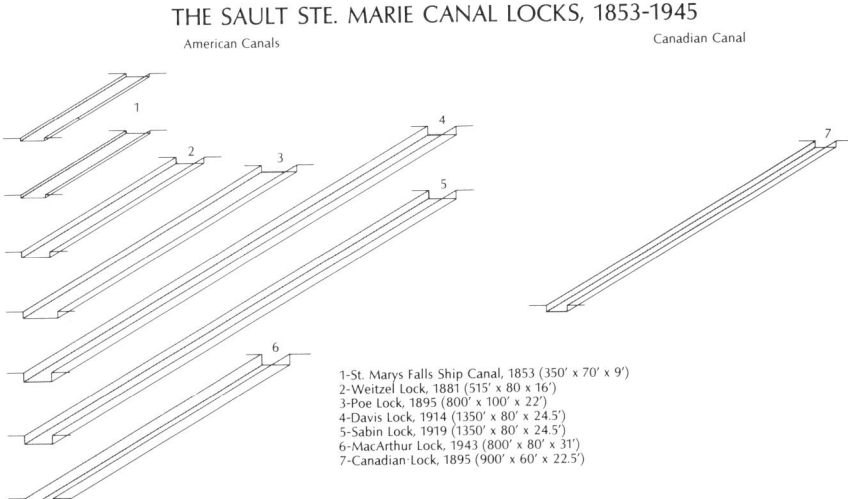

THE SAULT STE. MARIE CANAL LOCKS, 1853-1945

American Canals — Canadian Canal

1-St. Marys Falls Ship Canal, 1853 (350' x 70' x 9')
2-Weitzel Lock, 1881 (515' x 80 x 16')
3-Poe Lock, 1895 (800' x 100' x 22')
4-Davis Lock, 1914 (1350' x 80' x 24.5')
5-Sabin Lock, 1919 (1350' x 80' x 24.5')
6-MacArthur Lock, 1943 (800' x 80' x 31')
7-Canadian Lock, 1895 (900' x 60' x 22.5')

26 The Sault Ste. Marie Canal locks, 1853-1945. (Parks Canada)

The successive superintendents at the Canadian Sault knew they could lose this comparative advantage for the carriers were increasing in capacity, and the dimension most pressing the limits of the system was that of beam. The Americans were upgrading their facilities and so the canal staff at the Canadian Sault concentrated on the best means of improving the competitiveness of their operation. Moreover, the increasing volume of the freight also justified improvements in the system. The economic expansion of the day was producing a significant increase in traffic, which justified providing even better capacity at the Sault. The mineral deposits on the south shore of Lake Superior had long been developed, but by 1899 there was confidence that "the large iron ore and other mineral deposits on the north shore or Canadian side of Lake Superior are about to be developed in a marked degree."[44] Two years later these "mineral and timber resources" had attracted the promised attention and the 1901 report for the Canadian canal proudly recorded the passage of "the first cargo of Canadian iron ore ever shipped from the North Shore...carried by the steamer *Theano* of the Algoma Railway Steamship Line, bound from Michipicoten to Midland... with a cargo of 2,173 tons." These promising new developments together with the continued increase of traffic on the Canadian side to over 36 million tons by 1910 also caused some to argue for increasing the capacity of the Canadian system.

Several initiatives were pursued but the first was not advanced by canal interests. In the early 1900s, three railway companies applied for rights of way from the Sault Ste. Marie Bridge Company's tracks, across Whitefish Island, to terminals to be developed at the eastern end of the island at the foot of St. Mary's Falls. The island was still controlled by the Department of Indian Affairs and the companies requested the grant under section 5 of the Indian Act and sections 124 and 125 of the Railway Act by which such severances could be authorized if it were proven that the land was "necessary for Railway purposes." In 1900 the Algoma Central and Hudson Bay Railway applied for 11.4 acres across the centre of Whitefish Island, consent being given on February 12, 1900. In December 1901 the Pacific and Atlantic railway applied for the southern portion of the island amounting to 4.27 acres and were awarded it by order-in-council dated January 10, 1902. The only remaining land unassigned was the 3.65 acres of the northern part of the island which were applied for by the Ontario, Hudson's Bay and Western Railway Company in December 1901 and which were granted by an order-in-council of January 15, 1902.[45] The prime mover behind this development was F.H. Clergue, the dynamic entrepreneur and developer at Sault Ste. Marie. His ironworks were located above the locks and so his access to Lake Superior ore was direct and unimpeded; however, the coal required for his furnaces did have to negotiate the increasingly pressured lock system. Clergue was not unaware of the potential for power developments there either, but the submissions made reference only to the right of way for the three railways and terminals for the receipt of coal. Opposition to the project came from the Batchawana Bay Indians who still used the island, the town of Sault Ste. Marie which wanted the island developed for water power purposes and the C.P.R. who had their own plans for the as yet undeveloped property. Despite considerable opposition, letters patent to the three sections of land that make up Whitefish Island were issued to the three Clergue railways in February 1906. The portage railways could now proceed with their plans.

In the meanwhile, another approach was to make full use of the existing system, capitalizing on the differences in the design between the Canadian and Poe locks to the full advantage of the former. The two locks fought over inches. The Canadian lock's greater elevation allowed a 6-inch-deeper draught for up-bound vessels. The down-bound traffic encountered the same controlling elevations at both locks, but the Canadians turned to another stratagem to maintain their advantage: "an additional draft was obtained by opening the filling valves and raising the water in the lock chamber, which could not be done to the same extent in the Poe lock chamber owing to the additional width of forty feet over the Canadian lock."[46] Ross provides a fuller explanation of this imaginative use of his lock's capabilities in order to accommodate larger vessels:

> As the size of vessels navigating the Great Lakes has greatly increased, and a large number of them are built with a view of using the Canadian Lock to its full capacity, so far as beam and depth are concerned, it has become necessary in locking down-bound vessels, to partly open the valves behind the vessels so as to fill the void, as the vessel moves out. Hence has arisen the practice of what is called "flooding a vessel out." If this practice were not resorted to, the time consumed by a vessel in moving out of the lock would be increased from seven minutes to an hour or more....[47]

Additional capacity could be achieved at little cost by dredging the approach channels to allow vessels of deeper draught to make maximal use of the deeper Canadian lock. The increase in tonnage of 1.4 million tons in 1902 was thought to demonstrate that "the deepening of the lower channel has been appreciated by the vesselmen."[48] In 1903 contracts were let for the dredging of the upper channel to a depth of 21 feet 5 inches, "the same depth of water as is secured on the upper mitre sill of the lock" so that "a channel way will have been secured more in keeping with the large freighters now using it."[49] In 1910 engineer Fripp reported that since 1895 the draught of the upper channel had been increased about 4.5 feet to 21-22 feet deep, depending on lake levels; the upper approach had been dredged about 3 feet, providing a draught from 21 feet 5 inches to 24 feet 5 inches. After this dredging, therefore, the maximal draught of the lock was being used.

The only other option was the proposed construction of a larger lock facility. In 1906 Boyd opened his campaign for the upgrading of the Canadian lock:

> The day of the big boats has come, and there are now several building that have too much beam for passing through this lock, being 60 feet in width. The other day we locked through the Steamer *J. Pierpont Morgan*, being 600 feet in length and with 58 feet beam. We did this without any difficulty; this vessel was built by the steel trust, and was built with several others to fill the capacity of this lock, and not to be at the mercy of the American lock as to size in case of accident to that lock. In that

> case any of the 60 foot beam vessels caught above in Lake Superior would have to lie above the lock until the repairs were made before they could get through the locks as they would be too wide to come down through this lock. The question of a new and wider lock must soon be taken up by this department....[50]

Boyd had some concrete suggestions for his department, suggestions that also drew attention to developments which were threatening to limit the future development of improved facilities on the Canadian side:

> land lying between the present canal reserve and the rapids should be procured from the Crown Lands of Ontario before they are taken up by some private corporation so as to enable a new lock to be built, and by so doing obviate the trouble that the American government are now having across the river in obtaining lands required for their new lock, it having been taken up and built upon by private companies.

Others were concerned too, if only to protect their own interests. In 1909 the president of the Lake Superior Power Company wrote to the Honorable Frank Cochrane, whose provincial ministry of Lands, Forests and Mines controlled the islands in the rapids south of the canal, arguing that "These lands in the rapids should be reserved for public improvements in the interests of navigation...."[51] He noted that the St. Mary's River traffic had been increasing since 1855 and that the current trend would produce a volume of 100 million tons of traffic by 1915 and 200 million tons by 1925; further, he predicted that "practically the whole of the rapids on both sides of the international boundary will be required for navigation purposes." Although being more concerned with blocking the claims of other rival power developments to lands south of the Canadian canal, the case did highlight the need to reserve the lands there for future canal development.

In April 1910 engineer Fripp reported that having reviewed the proposed American developments, the advent of vessels 607 feet long, 60 feet in beam and 21 feet in draught, and the significant increase in traffic through the Canadian system over the previous 5 years, his reasons for the new canal were the following:

> First, to keep pace with the rapid increase in Canadian tonnage with the development of the west. Secondly, to provide a lock of width that will accommodate boats now plying on the lakes and being built of greater beam than the present lock will pass through. Thirdly, to provide additional draft both in the lock and approaches so that vessels may load to pass through the Canadian canal drawing as much water as will be provided by the new canal now under construction by the United States government.[52]

Fripp reported that surveys had been started "for a proposed new ship canal," three alternative lines had been laid, and data was being gathered "to form an approximate estimate of the cost of a lock that will meet the requirements of the estimated increase and development of lake commerce." Apart from developing the appropriate design for the new project, an essential prerequisite was that the necessary lands be acquired to accommodate the new locks. Despite Boyd's warnings, the lands south of the existing canal, in particular Whitefish Island, had been patented and were, therefore, controlled by others. Even this obstacle was removed, however, after the government's expropriation of the lands in September 1913. *The Sault Daily Star* (Sept. 25, 1913) trumpeted the news of a $20-million expenditure on a new canal at the Sault. The headlines blazoned the news of the recent developments:

> PROPERTY EXPROPRIATED YESTERDAY FOR THE NEW CANADIAN SHIP CANAL PROMISED BY MINISTER COCHRANE
>
> GOVERNMENT ASSUMES CONTROL OF PROPERTY SOUTH OF PRESENT LOCK TO INTERNATIONAL LINE

Frank Cochrane, who represented northern Ontario in R.L. Borden's government after 1911, was interested in all developments in his region. His announcement confirmed that the government was considering upgrading the locks at the Sault to keep pace with the improvements being contemplated for other sections of the system such as the Welland. While not prepared to comment on the decision to construct, the local M.P., A.G. Boyce, did offer his opinions on the matter, returning to Tupper's "National" argument of a generation earlier:

> Canada requires a strictly national waterway from the lakes to the sea, subject only to the control of our own Government, free from all the problems connected with the development of International waterpowers, and the regulation and control of international waters secure from any diversion of traffic to a foreign country en route to the seaboard, ensuring the export trade of Canadian products to Canadian seaports, and such as will best serve to build up inter-provincial traffic in the commodities which are suitable for water carriage, viz. ores, iron and steel, coal, pulpwood, pulp, paper, lumber and grain, and which form 97 per cent of the traffic of the Great Lakes.

Despite the fragility of some of these arguments, Boyce's views were well received at Sault Ste. Marie and gave some hope for future developments there. Further news of the proposal leaked out of Ottawa on September 25, 1913, and was published in newspapers throughout Canada and the United States. The Toronto *Globe* commented on Cochrane's "big scheme" September 26, 1913:

> The enlargement of the Soo Canal is a logical sequence to the enlargement of the Welland Canal. The next step will be the enlargement of the whole St. Lawrence canal system so as to give a continuous thirty-foot channel right through from the head of the lakes to the Atlantic. That is the big scheme which Hon. Frank Cochrane is reported to have in view, and which the Government it is said will adopt. The formal announcement of the whole scheme will probably not be made for some considerable time yet, but meanwhile the enlargement of the Welland and the Soo Canals will be proceeded with.

The article went on to recognize that others were advancing the case for the Georgian Bay—Ottawa River Canal route as an alternative to the St. Lawrence, but concluded that "Meanwhile the enlargement of the Soo Canal to a depth of thirty-one will not interfere with the dangling of both projects before the eyes of the electors at the next general election." Political cynicism aside, there were other reasons for considering that the discussion of the proposed project was nothing more than an exercise in pre-election politicking. The new American canals at the Sault were well under way and were to be operational in 1914 and 1919. Even the most optimistic estimates for the future increase in traffic could be accommodated in the immediate future by these new developments. Further investments at the Canadian Sault would not interfere with the current split in the grain flows to the Georgian Bay and Buffalo railheads, but improvements of the Welland and St. Lawrence navigation might better serve Canadian needs. Despite the flurry of activity in 1913, therefore, the Canadian lock was not upgraded or replaced. However, the government had reasserted its control over the lands that had been lost temporarily to private interests and retained them for possible future developments in the national interest be it navigation, rail transport or water power.

In 1914 and 1919 the Americans opened two new locks, the Davis and Sabin respectively, each being 1350 feet by 80 feet by 24.5 feet. In 1914 the largest vessels on the lakes were 625 feet long, the new locks being capable of locking through two of them in one lockage. The Canadian superiority in lock draught had been surpassed, but Ross continued to be sanguine about the competitive advantage of his canal. Referring to the dimensions of the Davis lock, or "Third Lock" as it was referred to popularly by the Americans at the time, Ross argued "While this depth is much greater than that in either the Poe lock or Canadian lock, the increased depth cannot be used until the depth in the river reaches is made to correspond." On the Canadian side, however, the dredging of the approaches had been sufficiently effective that "the depth of water in the Canadian lock governs the loading draught of vessels."[53] The traffic returns of the following year dashed these hopes. The opening of the "Third Lock" had produced a "great falling off in the traffic through the Canadian canal" from 27.6 million tons to 7.8 million tons of freight. The equivalent figures for the American system showed an increase from 27.8 million tons to 63.6 million tons. The lesson was obvious:

> Before the opening of the new lock the Canadian canal had an advantage of 6 inches in draught over the American canal, and was the point in the system of navigation which governed the loading of vessels. Since the opening of the new lock, the Canadian lock has lost this advantage in draught, and the point which governs the loading draught of vessels is somewhere in the St. Clair river.[54]

After the completion of the "Fourth," or Sabin lock, on the American side in 1919, there were two locks in service there longer and deeper than their Canadian rival. Accordingly, the Canadian traffic declined even further and by 1921 had dropped to below 2.0 million tons of freight. After the replacement in 1942 of the Weitzel lock by the even larger MacArthur lock (capacity 800 feet by 80 feet by 31 feet), the American system was clearly dominant. By 1945, therefore, the St. Mary's Rapids had been bypassed by seven canal schemes, each progressively larger and more efficient than the previous one, and the Great Lakes navigation system was served there by five efficient locks.

The Development of the Lake Fleet

In the analysis of canal traffic, therefore, the size of the carriers in relation to the capacity of the locks becomes the critical factor in determining the utility of the lock systems available.[55] Several indicators of the capacity of the vessels using the locks may be used. The overall dimensions of length, beam and draught pose obvious limitations on the use of locks at any time, but one of these was somewhat adaptable; processes of "lightening cargo" by the vessels and of "flooding through" by lockmasters allowed some relaxing of the draught constraint. Restrictions posed by length and breadth were, however, insurmountable. The other definition of size of vessels, that of tonnage, is more confusing, several measures being used.[56] "Deadweight" refers to the total carrying capacity including cargo, fuel, fresh water, stores, etc. "Net Registered Tonnage" is the total internal cubic capacity of the ship (100 cubic feet equalling 1-ton capacity) minus spaces devoted to non-cargo activities such as crew, fuel and engines. The constant increases in deadweight and registered tonnage were always accompanied by increases in overall dimensions with implications for the canal systems being used.

When the Canadian Sault first opened its gates to the lake carriers, few exceeded 1000 tonnage, 200 feet long, 40 feet beam and 20 feet draught. Indeed, the great majority of vessels were much smaller, including a miscellany of schooners, steamers and small barges. During the 1896 season the average tonnage of the 4560 vessels passing through the American and Canadian canals was a modest 810 tons. In 1898 the Sault canals recorded 17 761 passages of which 3431 were by 78 craft (an average of 44 passages each) and an average registered tonnage of a mere 31 tons. Further, of the 856 registered vessels using the canals in that year, 333 were sail and 523 were steam.[57] Larger vessels were demanded by the growing trade in ore and grain and one new class of larger freighter was the distinctive group known as "whalebacks," some 40 of which were built at Duluth, Minnesota, between 1888 and 1898. Typical of one of these unique vessels is the *J.B. Neilson,* a 2234-ton vessel 308 feet by 42 feet by 25 feet. Even these were soon to be surpassed. Writing in 1898, Superintendent Boyd looked back over the 1897 navigation season and predicted even larger carriers in the near future:

> This season [1897] the record-breaking cargo has gone over the 7,000 ton mark, and it is expected that before the snow flies that some of the larger ones now on the stocks will have equalled if not extended beyond the 8,000 ton mark. These larger carriers are forcing the smaller ones to be laid up, and anything smaller than 2,000 tons except for lumber and some local purposes will soon be things of the past.[58]

By the turn of the century, vessels had increased considerably both in overall size and in carrying capacity. By the late 1890s the largest vessel on the Great Lakes was the 4477-ton *B. Morse* having a 7500-ton capacity for carrying ore. Her 456-foot length, 50-foot beam, and

29-foot draught ushered in a new era of cargo carriers. The review of the 1898 traffic through the Sault recognized the import of such developments on the direction and increasing volume of traffic:

> the apparent decrease in the percentage of [Canadian traffic] is from the building of large American vessels, there being at present quite a number over 470 feet long and some now building and to be ready for next season's work are to be 500 feet in length. The 8,000 ton cargo is now of the past it having been broken by two vessels. The first of this large class of vessels to pass through these canals was the schooner *Manilla* of the Minnesota steamship line, with a cargo of 8,250 net tons of iron ore on a draught of 18 feet 1 inch, followed after by the schooner *John Smeaton* of the Bessemer steamship line with a cargo of 8,339 net tons of ore on a draught of 18 feet, both of these cargoes passed down through the American lock. The largest cargo carried through the Canadian lock is to the credit of the steamer *Henry Oliver*, 476 feet long, with a net registered tonnage of 3,617 tons, drawing 18 feet of water and carrying 7,464 net tons of ore.[59]

In 1900 four other 8000-ton cargo carriers, the *John W. Gates, James J. Hill, Isaac L. Ellwood* and *William Edenborn*, came into operation, each one 478 feet by 52 feet by 30 feet. The Sault superintendent recognized the implication of this for his 900-foot-long lock, designed for locking two or more ships simultaneously. The increase in beam size was also beginning to test the capacity of the Canadian lock:

> The day of the large and deep draft vessel is to hand there being now some 7 or eight of the 500 foot class with 52 feet beam. As it is now there are several steamers towing schooners that neither the big American Lock nor this one can accommodate the two at the same time. There is strong talk and very strong pressure being brought to bear upon the American congress to have a new Lock built on the site of their present old lock and it is to be some 1310 feet long and over 100 feet wide so as to be able to take in 4 of the 600 footers, if they are ever built, as it was supposed that the present large lock (called the Poe lock) would when built be large enough to take at one time four of the largest boats on the lakes for a long time to come, whilst now it cannot take in two of them, and this within four years after its completion.[60]

Within the decade the first of the 600 footers, the *J. Pierpont Morgan* entered onto the lakes. Launched in 1906, she was 580 feet by 58 feet by 32 feet, over 7000 registered tons and a deadweight of about 13 000 tons. The size of cargoes locked through the canals continued to increase, the record cargo of 13 978 tons passing through the Canadian Sault on September 12, 1908. No one could say that Boyd had not warned the Department of Railways and Canals of these developments and their impact on the Canadian system. The *J. Pierpont Morgan* was soon followed by eight freighters of similar capacity, all with a 58-foot beam allowing the critical 2-foot clearance for the Canadian canal. By the early 1930s some 30 vessels of this type were being operated by the "Steel Trust" fleet of the Pittsburgh Steamship company alone.

Whereas in 1884 the Great Lakes fleet could boast only 18 vessels carrying cargoes of more than 3000 tons, with none exceeding 4000-ton loads, by 1906 the situation was much different:

Tonnage	No. vessels
0-3 000	454
3-4 000	134
4-5 000	37
5-6 000	24
6-7 000	52
7-8 000	82
8-10 000	48
10-12 000	36
12-14 000	12

The breakdown of the vessels by their overall dimensions reveals a similar pattern:[61]

Length	Beam	No. vessels
0-300 ft	0-38 ft	523
3-400 ft	38-50 ft	171
4-500 ft	45-53 ft	128
5-600 ft	52-60 ft	57

Boyd's forewarnings about the 60 footers had been realized in 1907 with the launching of the "Steel Trust's" *D.M. Clemson,* 580 feet long, 60 feet beam and 32 feet draught. Twelve others were added to the fleet over the next 20 years, which meant that whereas the vessels of deeper draught had occasionally used the Canadian lock when travelling empty or with lighter loads, the beam of these newer vessels excluded them entirely.

A new level of operation was attained by the *Harry Coulby* in 1927 with her delivery of a cargo of 14 000 tons of iron ore to Cleveland, Ohio. Or again, the Canada Steamship Line's *Lemoyne;* built in 1926 and over 600 feet long, she proved both her capacity and diversity by carrying a cargo of 16 577 tons of coal in May 1928 and another of 571 960 bushels of wheat the following July. Other vessels of the Canada Steamship Line fleet of this dimension included the *Ashcroft, Gleneagles* and *Stadcona.* All close to 600 feet long and 32 feet draught, more critically they were 60 feet in beam with the *Lemoyne* being some 70 feet wide. The implications for the Canadian Sault traffic were clear. Record cargoes such as those of the *Lemoyne* and her sister ships were not to be matched until the World War II development of vessels with deadweights of 20 000-25 000 tons.

While these large vessels were being developed to accommodate the ever-increasing volumes of iron ore, wheat and timber passing through the canal, numerous smaller vessels continued to ply their trade, some as bulk freighters and others as package carriers. Known as "canallers," they carried grain, coal, steel and general cargo the length of the Great Lakes—St. Lawrence navigation system. Generally smaller in size, they navigated the lakes, negotiated the canals, and entered the river sections of the systems without the necessity for "breaking bulk." In the first quarter of the 20th century, the largest fleet of these vessels was that of the Canada Steamship Lines, consisting of more than 100 vessels. Both the package freighters and bulk cargo canallers were generally under 2000 registered tonnage, under 250 feet long, less than 45 feet beam and less than 25 feet draught. Ships like these were the real workhorses of the Great Lakes system and constituted the greatest number of lockages through the canals. Throughout the 1895-1945 period of operation of the Canadian Sault, the average size of vessel locked through exceeded 3000 tons only in the exceptional years of 1912 and 1913. In all other years the average seldom exceeded 1500 tons, indicating the Canadian Sault was being both avoided by the larger vessels and favoured by the smaller ones.

Finally, perhaps the most popular lake carriers and the ones reminiscent of a world gone by were the grand passenger vessels operating between the lower lakes and the lakehead. Three of the first to use the Sault canal were the "triplets," the *India, China* and *Japan,* which ran from Buffalo to Duluth. Built in 1871, the three vessels were 210 feet long, 32.6 feet beam and 14 feet draught. Their speed was 12 m.p.h. and their itinerary was similar to the following one of the *Japan* in 1896:[62]

	ARRIVE	**DEPART**	
Buffalo		1 PM	Thursday
Erie	9 PM	11 PM	
Cleveland	8 AM	8 PM	Friday
Detroit	7 AM	Noon	Saturday
Mackinac Island	1 PM	8 PM	Sunday
Sault Ste. Marie	8 AM	Noon	Monday
Marquette	10 PM	1 PM	
Hancock	10 AM	3 PM	Tuesday
Duluth	8 PM	9 PM	Thursday
Hancock	Noon	5 PM	Friday
Marquette	11 PM	10 AM	Saturday
Sault Ste. Marie	8 AM	9 AM	Sunday
Mackinac Island	3 PM	5 PM	
Detroit	6 PM	9 PM	Monday
Cleveland	8 AM	10 AM	Tuesday
Erie	9 PM	11 PM	
Buffalo	8 AM		Wednesday

The C.P. ran three sister ships in the passenger trade, the *Alberta, Algoma* and *Athabasca.* Completed in 1884, these vessels had been built in Britain, sailed across the Atlantic, each cut in two to negotiate the St. Lawrence canals, and reassembled at Buffalo for the Great Lakes trade. All three had the same dimensions: 2282 tons, 263 feet long, 38 feet beam and 23 feet draught. In 1907 the fleet was expanded with the addition of two new and larger vessels, the *Keewatin* and *Assiniboia,* and the company embarked on its peak period of operation in the passenger and freight trade. Its original eastern terminus at Owen Sound was moved to the new C.P. facilities at Port McNicoll in 1912, the western terminus being Fort William. Another major passenger carrier was the Northern Navigation Division of the Canada Steamship Lines which in the 1930s ran three passenger vessels from Windsor, Detroit and Sarnia to the head of the lakes, with stops at Sault Ste. Marie, Port Arthur, Fort William and Duluth on the 7-day voyage. The *Hamonic, Huronic* and *Noronic* were about 5000, 3000 and 7000 tons respectively, less than 400 feet long, about 50 feet beam and 30 feet draught.

The changing technology of canal transport, in association with the increased size of lake carriers, resulted in differences in the degree of use made of the Canadian and American canals respectively. When first constructed, the 22-foot-draught Canadian canal held an advantage of several feet over the St. Mary's Falls (9 feet) and Weitzel (16 feet) canals and even an advantage over the new Poe canal, if only measured in inches. This ended with the opening of the 24.5-foot-draught Davis and Sabin locks in 1914 and 1919 respectively, and the 31-foot-draught MacArthur lock in 1943. Differences in length and breadth were also to become significant and even the Canadian Sault's record-breaking 900-feet-by-60-feet dimensions were soon to prove inadequate for some of the leviathan carriers being developed. While the American system was continually expanded to keep pace with these developments in Great Lakes marine engineering, the Canadian system continued to service vessels of late-19th-century dimensions. Both the extent and constitution of its traffic reflects this fact.

Lockages, Freight Tonnages and Passengers

The very nature of the basic economic geography of the continent predicated that there would be a flow of traffic through the channel at the Sault. As with the fur trade, western staples such as iron, grain and lumber flowed east to consumers and processors in both eastern American and overseas industrial-urban markets. Manufactured commodities and more exotic or specialized resources moved west, but always in more restricted flows. The eastward flow of staples dominated the traffic.

The Sault canals were initially built for the iron ore trade and from the very first years of operation in the 1850s, iron ore dominated the traffic there. The ore moved from the mines of the Lake Superior range at Mesabi, Menominee, Gogebic, Marquette and other locations to such ports as Duluth, Two Harbours, Ashland, and Escanaba. Once loaded on the lake carriers, it moved east through the Sault locks.

Indirectly, this also generated a return flow of commodities. The very presence of the local ore, together with the economic benefits from break-in-bulk advantages, dictated there would be iron production at such locations as Duluth and Sault Ste. Marie. The fact that the coal required by these plants was located in such eastern states as Pennsylvania, West Virginia, Kentucky, Tennessee and Illinois necessitated a compensating return flow of coal from east to west.

The late 19th century was an era of rapid industrialization and urbanization in the East and settlement expansion in the West. The traffic through the Sault canals reflects this. The burgeoning production of grain flowed from the western "bread baskets" to the tables of the urban population in eastern North America and overseas. The grain trade began in the 1870s and was focussed on Duluth. The construction of grain elevators and shipping terminals at Fort William and Port Arthur led to the "Lakehead" dominating the western terminus of the Great lakes grain traffic to the same degree that Buffalo dominated the eastern terminus. And it all passed through the Sault. The increasing industrial plants of the eastern industrial belt required the iron as much as the proliferating newspapers and construction industries required the pulp and lumber respectively. The traffic at the Sault tells a story about the economic history and economic organization of the continent during the first half of the 20th century. Several dimensions of this traffic merit particular attention: the volume, direction and constitution of flows.[63]

The first noteworthy characteristic concerning the volume of the Sault canal system is the constant increase in volume of traffic and vessels using the system. During the 19th century this phenomenon invited comparison with that of the then more famous Suez Canal while by the 20th century the Canadian and American canals at the Sault had surpassed the volume of the Suez, Panama and Kiel canals combined. More specifically, the Sault canals accounted for the following totals between 1895 and 1945:

Freight Tonnage in Millions of Tons

	Canadian	American	Total
1895	0.75	14.47	15.22
1900	2.04	23.61	25.65
1905	5.47	38.80	44.27
1910	36.40	26.00	62.40
1915	7.75	63.55	71.30
1920	2.48	76.80	79.28
1925	1.63	80.24	81.87
1930	1.69	71.21	72.90
1935	1.93	46.36	48.29
1940	1.96	87.90	89.86
1945	2.02	111.26	113.28

Although the total system experienced growth, it was not distributed equally between the two national systems. Clearly, the traffic through the Sault canals was dominated by the American locks. Between 1906 and 1914, however, the trend reversed; the Canadian traffic increased annually to some 36.4 million tons in 1910 and peaked at 42.7 million tons in 1913. During the same period, the American traffic declined, falling to a low of 22.5 million tons of freight in 1911, but this reversal was as artificial as it was short-lived. It reflected the interruption in the traffic through the American system caused by the construction of the new Davis and Sabin locks. On their completion, Canadian traffic dropped dramatically. When the Davis lock became operational in 1914, the Canadian Sault's traffic tumbled from its peak of 42.7 million tons in 1913 to 27.6 million tons in 1914 and fell even further to 7.8 million tons in 1915. The work on the Sabin lock again diverted some traffic through the Canadian locks, with freight tonnages increasing to 16.8 million tons in 1916. On the opening of the new lock in 1919 the volume fell to 4.2 million tons in 1919 and to 2.0 million tons by 1921, a level of activity that, apart from a brief surge to 4.6 million tons in 1943, became characteristic for the rest of the period to 1945. The American system, however, with its new capacity installed and operational, experienced growth over the same period. By 1923 the volume of freight approached 90 million tons, dropped off during the Depression to a low of 18 million tons in 1932, peaked at 114.5 million tons in 1942, and closed the period with 111.3 million tons of traffic in 1945.

Another dimension of the traffic flow does not reveal the same degree of decline relative to the volume of American canal traffic. The number of vessels locked through the Canadian Sault climbed steadily to a peak of 8285 vessels in 1913 at which time 18 599 vessels used the American canals. Thereafter, in keeping with the greater use of the American system, the number of vessels through the Canadian lock declined constantly through the 1920s and 1930s. The records revealed a short-lived upsurge during the war years of 1939-45, but nevertheless the 2678 vessels passing through the Canadian Sault in 1945 constituted the lowest traffic volume since the canal opened in 1895. Whereas the decline in tonnage of freight handled had been to less than a fiftieth of the Sault system, the Canadian lock still handled one-eighth of the lockages of vessels. This disparity reflected several factors. Smaller vessels still operated on the Great Lakes and those vessels favoured the speedier lockage through the Canadian lock; apart from a minor upswing to over 3000 tons in 1912, the average tonnage of the vessels using the Canadian lock generally ranged between 1000 and 1500 tons while that of the American system increased constantly, reaching 3000 in 1916, 4000 in 1928, and closing the period with an average of 4176 tons for its 19 814 vessels in 1945. Even some larger vessels favoured the Canadian lock under certain circumstances. Noting that in 1922 there had been an increase in registered tonnage through his lock at the same time as there had been a decrease in freight, Ross explained "owing to the deeper drafts of the American canals the loaded vessels are compelled to go that way while the light [unloaded] vessels are using the Canadian canal."[64] Finally, as will be seen, the majority of the passenger vessels used the Canadian lock.

Not only does the volume of freight and registered tonnage demonstrate the diminished, if still significant, role of the Canadian lock in the system at the Sault, it also underscores the erosion of the Sault lock's initial prominence in the total Canadian canal system. The following table compares the tonnage of freight handled by the major canal systems in Canada for selected years, the category "Others" being made up of the relatively minor elements of Chambly, Ottawa, Rideau, St. Peters, Trent and Murray.

Comparison of Total Freight in Millions of Tons

	1896	1913	1925	1945
Welland	1.28	3.57	5.64	12.04
St. Lawrence	1.11	4.30	6.21	4.36
Sault Ste. Marie	4.58	42.70	1.63	1.87
Others	1.02	1.48	0.65	0.32
Total	7.99	52.05	14.13	18.59

The Sault had slipped from the position of dominance it held in the Canadian system for the first two decades of its operation when it accounted for some 57% of the total freight carried in 1896 and the astronomical 82% in the peak year of 1913. The development of superior facilities at the American Sault and the increased importance of the Welland led to a decline in the relative position of the Canadian Sault to a modest 10% of the volume of Canadian canal traffic in 1945.

The composition and direction of the flows of commodities moving through the Sault may best be treated together as they are closely linked. The salient feature of the traffic pattern is the predominance of the "down" or eastward movement over the "up" or western movement of total freight. This was true for both the Canadian and the American locks. In the very first year of operation, the Canadian lock passed 415 432 tons "down" and only 180 405 tons up. Apart from this pronounced imbalance, even more striking was the dominance of certain commodities in the respective flows. Three cargoes dominated the "down" movement: iron ore (52%), wheat (33%) and lumber (5%). Moreover, coal constituted 82% of the "up" traffic. This pattern was to continue and even become more firmly established over the next decade. The trade figures for the Canadian lock in 1907 show that of the 15.6 million tons handled by the lock, only 19% constituted "up" traffic, the great bulk of the traffic being the 12.6 million tons moving "down" to the eastern cities and industries. As had become established, this "down" freight was dominated by iron ore (79%) and wheat (12%). Even more pronounced was the presence of coal (81%) in the "up" cargoes. The peak year of 1913 with its 42.7 million tons of freight handled by the Canadian lock was no different; iron and wheat moving east accounted for 76% and 9% respectively, while coal from the Lake Erie coal ports accounted for almost 9% of the total traffic and a staggering 75% of the "up" cargoes.

Following the assertion of the dominance of the American locks in the big-carrier trade, the traffic of the Canadian lock became more general and less specialized. The largest vessels introduced in the 1930s and 1940s were the leviathan ore and grain carriers, none of which could negotiate the Canadian lock with its limited 60-foot width and nominal 22-foot depth. By 1934 the combination of the economic depression and these changes in navigation caused a decrease in the former dominants in the trade. In that year no iron ore moved through the Canadian lock. The war years occasioned a minor increase in that cargo, but by 1945 only 600 tons of the over 2 million tons moved in that year are iron ore. Coal continued to dominate the "up" trade in the 1930s and 1940s and amounts to about 6% of the total trade in 1934 and 1945. The wheat flows are also affected by the increase in size in carriers and only 631 306 (37%) tons passed through the Canadian lock in 1934, falling to 590 562 bushels in 1945 (29%). The effect of the American locks was being felt. Whereas in 1914 over 81% of the Canadian wheat moving east passed through the Canadian locks, by 1926 a mere 3.9% passed through there, the rest, over 96% of it, being carried through the Davis or Sabin locks in such giant carriers as the new *Lemoyne*. The pattern was set for the next decade.

If the fortunes of the Canadian lock appear to fall under the shadow of its American competitors after 1914, one aspect of the Great Lakes traffic continued to favour the Canadian Sault. The competition for passenger traffic was clearly won by the Canadian Sault:

	Canadian locks	American locks
1895	2 326	30 910
1900	22 280	36 313
1905	26 147	28 401
1910	33 609	33 324
1915	25 047	25 378
1920	43 455	24 999
1925	34 743	22 352
1930	27 831	17 627
1935	19 558	13 352
1940	31 664	21 143
1945	39 247	6 797

In this category of cargo the trend is clear. The passenger traffic peaked in 1911, the year when some 80 000 passengers were carried through the Sault system, 40 245 of them through the American Weitzel and Poe locks. From that peak year, the American passenger traffic declined rapidly to a mere 529 passengers in 1942 and the low of 55 passengers for 1944. Conversely, the Canadian traffic flourished during this period. In 1920 some 43 455 passengers were locked through the Canadian lock although traffic decreased to the low of 11 193 passengers in the depression year of 1933. Generally throughout the 1920s and 1930s, however, there appears to have been a vibrant passenger trade with figures in the 20 000 range. The peak year was reached in 1941 with 57 783 persons passing through the Canadian lock at the Sault.

The reasons for this dominance are varied. Certainly, the Canadian Pacific line and the Northern Navigation Division of the Canada Steamship Line dominated the passenger carrying trade in the Upper Lakes. Their dimensions were suited to the Canadian lock and they were not obliged to compete with the heavy traffic which concentrated on the American locks. During World War I, not only did the wartime demand for raw materials generate record traffic, but *The Sault Daily Star* (Dec. 28, 1916) reported that "Owing to the war, the summer tourists, who have had Europe barred from them and who are instead seeing America, have increased the number of passengers by 4,586 or 9 per cent."

Tourism continued to flourish after the war. On August 23, 1919, a headline in *The Sault Daily Star* boasted that "Tourist Traffic Through Sault Makes a Record: Every Passenger Vessel Passing There is Crowded These Days"; it continued, that "The locks which lead to one of Canada's most wonderful waterways, the wildly beautiful scenery on the shores of Lake Superior, are probably the main attractions to sightseers, many of whom take the trip yearly." A later report (July 12, 1921) claimed that "Train Traffic Falls Off as Lake Route Proves Popular," advancing the explanation that the "cool lake breezes" afforded greater comfort to travellers during the summer months. An article in *The Sault Daily Star* (Aug. 21, 1913) reported that "Tourist Traffic This Season Has Been Very Large," and that "people outside are now beginning to realize that the St. Mary's Rapids at Sault Ste. Marie is one of the few places where that delight of the fisherman, the rainbow trout, can be caught." The town's other attraction, the ship canal, was brought to the nation's attention when it was used as the motif for the $4 bill in 1902; this was a correction because an earlier representation of this national achievement on the 1900 Dominion of Canada $4 bill had been of the canal at the American "Soo."

Whatever the reason, the passenger traffic on the Great Lakes and through the Canadian Sault was a major feature of the society of the day. Having ports of call at the major settlements of the north shore, and good connections with rail termini, the Upper Lake passenger boats served the public need for regional transport and recreation long after other traffic had deserted the Canadian lock.

The "Soo" as a System

From the outset of the construction of the Sault locks, the Americans held the lead. Motivated by an earlier western development, energized by an industrialization that was ahead of that of Canada by some three decades, and supported by vaster resources, the American momentum for improvements at the rapids at St. Mary's River was always strong. The Americans never competed with the Canadians there. It was simply that a system had to be provided. Indeed, once Canada had its own nationally controlled lock which gave the semblance of independence from foreign control of navigation, the need for further independent development lessened. Certainly, once the Davis and Sabin locks had been added to the Poe and the Canadian, an additional development would have been surplus capacity. It was better that Canada direct its resources to the Welland and the St. Lawrence than invest further in a system that could only increase American traffic, much of which was directed to the American heartland.

It is somewhat ironic that the Canadian canal worked hardest and served the continental needs best when called to act as an accommodation for the flows blocked during the construction of additional American capacity. It also provided a useful alternative route for smaller cargo vessels, for others returning "up" under ballast, or as an appropriate system for the passenger traffic. From the outset, however, it was of secondary importance to the locks developed "across the line."

V EPILOGUE: THE RAPIDS, THE CANAL AND THE TOWN

The "Soo" has had several connotations for the Canadian population. To some it was one of the great portage routes for the fur trade and it has, therefore, an important place in Canadian history. To others it is the Sault canal, one of a series of American and Canadian canals, that figures large in the litany of late-19th-century technological accomplishments intended to further national development and identity. To still others it is the town, an industrial town of iron and steel works and lumber and wood products, that symbolizes the Canadian industrial establishment supported by the resources and electrical power of Ontario's north.

All three images, the rapids, the canal and the community of Sault Ste. Marie, are closely integrated into the image and reality of the "Soo." The canal has been but one part of the economic and social experience of the town, albeit a primary one.

From "Suburbs" to Municipality

In his novel *The Rapids,* Alan Sullivan described the bucolic settlement of St. Mary's before the arrival of the entrepreneurial adventurers who were to transform the natural resources of the region into industrial products and the community into a city:

> There may be communities now such as St. Marys was twenty-five years ago, but one goes far to find them. Electricity has altered their distinctive character. The traffic of half a continent glided majestically past these wooded shores, with the deep blast of whistles as the great vessels edged gingerly into the Government lock across the river to be lifted to Superior, and another farewell blast as they pushed slowly out, and lastly a trail of vanishing black smoke as they dwindled westward to the inland sea. For seven months this procession passed the town but never halted, till the people of St. Marys felt like the farmer who, in mid field, waves a friendly hand to a speeding train....There was no telephone to jangle, no electric light and no water-works, but in the soil of St. Mary's were springs of sweet water, and through the windows came the soft glow of lamplight as evening closed in, and the shuffle of feet on the porch announced the visitor. It was from the river and the close encircling forest that St. Marys took on its atmosphere. The maple bush was full of game, and the beaver built their curved dams in tamarac thickets within three miles of the village. It was a common thing to kill Sunday's dinner in a two hours stroll from Filmer's store, and at the foot of the rapids where the Indians pushed their long canoes up to the edge of the white water, there were great, silver fish for the taking....All this and

much more had the folk of the village, and everything that went to make up a sweet, clean, uneventful life. And then into this Arcadia dropped one day a stranger, with an amazing experience of the outer world, a kaleidoscopic brain, an extra-ordinary personal magnetism and a unique combination of driving force and superlative ambition.[1]

Sullivan's St. Mary's is unmistakably Sault Ste. Marie and he has captured the essence of the preindustrial community there and also identified imaginatively the elements of change. Railways, electricity, telephones and "strangers" with magnetism and drive were central to the waves of "progress" sweeping into "Sleepy Hollows" throughout North America in the latter part of the 19th century. Sault Ste. Marie had all these ingredients.

The official pronouncements were prosaic statements of the same principles. Thus, in 1887 the residents and ratepayers of Sault Ste. Marie petitioned for the incorporation of their town, arguing "it is expected that the lands hereinafter described will rapidly increase in population upon the construction of a line of railway to them, and that such line of railway will shortly be completed, and that various manufactories will utilize the unimproved water power included in their limits."[2] Although not as emotive as Sullivan's prose, this "bureaucratese" did identify some of the same portents of population growth and economic development: railways, electricity and industry. These were the forces of change to be marshalled by the contemporary barons of industry.

Sault Ste. Marie's growth is demonstrated best by the census return for the community between 1871 and 1951:[3]

Year	Population	% Change
1871	879	2.1
1881	780	-11.3
1891	2 414	209.0
1901	7 169	197.0
1911	10 984	53.2
1921	21 092	92.0
1931	23 082	19.4
1941	25 794	11.7
1951	32 452	25.8

In 1881 the Sault settlement was stagnant, if not declining, despite the promise of development of northern forest and mineral resources. The 1880s saw the onset of the great economic boom and Sault Ste. Marie's population grew accordingly. By 1901 it had grown tenfold and doubled to over 21 000 by the 1920s. While the rate slackened somewhat, growth continued at about 1% per year, another upswing occurring during the forties when the war stimulated local industries.

Another facet of the demographic changes experienced by the town during this period was the emergence of a more polyglot population. While this was a reflection of the increasingly diverse immigration to Canada in the late 19th and early 20th centuries, the concentration of large numbers of "new immigrants" at the Sault reflected the remarkable opportunities for employment. Apart from the government ship canal and the Lake Superior Power Canal, the various mineral, lumber and transport enterprises of the Clergue industrial empire were all developing during this period. The British and French were joined by less familiar faces and accents and while many moved on following the completion of the projects, many others stayed to become permanent residents of the growing community:

Ethnic Origins of Sault Ste. Marie's Population[4]

Year	British	French	Italian	Others
1871	46.9%	32.5%	-	20.6%
1881	72.4%	19.2%	-	8.4%
1891	-	-	-	-
1901	71.7%	10.2%	7.3%	10.8%
1911	64.1%	9.6%	11.0%	15.3%
1921	61.2%	10.5%	13.1%	15.8%
1931	57.6%	9.3%	14.1%	19.0%
1941	57.8%	11.0%	14.0%	17.2%
1951	55.4%	12.7%	13.4%	18.5%

Of the one-third of the population who were neither British nor French, the Italians were the predominant group while the Finns, Ukranians and Scandinavians constituted other prominent ethnic groups in the community. Apart from their cultural distinctiveness, their presence was all the more obvious by their concentration in the "West End" of the town, close to the canal, the railway and the industries which had brought them to the area. Not all stayed. *The Sault Daily Star* of March 5, 1903, reported there were only 1500 Italians in "Little Italy" at that date, about 3000 others having left since the previous year. As welcome as their labour had been during the construction years of the canal and the Clergue empire, their presence as fellow citizens was not appreciated by the Sault establishment. Memories of past labour strife, together with the endemic problem of the fear of "the strangers within our gates" (as J.S. Woodsworth titled his famous 1908 book on immigration), produced considerable tension and hostility. Inflamatory *Star* editorials increased hostility: "A Profitable Change" (Oct. 20, 1910) was an overt attack on the Italian population, commenting "What a great difference it would make in the Sault to replace these aliens by English workmen...." Eventually, such tensions were eased, but they were very much part of the social structure developed by the rapid economic growth of the community.

Several factors favoured the growth of the Sault and the canal was but one of them. Its construction contributed to the "boom" atmosphere of the 1890s. The eventual operation and functioning of the Canadian Sault ship canal was an essential element of the economic infrastructure of the town and region following its completion in 1895.

Economic Links

Transport costs are an important factor affecting the location of economic enterprises. Because water transport affords the cheapest means of transporting bulk cargoes of low unit value, heavy industry tends to locate on water transport routes. Another important locational principle is that where any form of transhipment occurs, there are economies to be derived from locating processing and fabricating industries at such break-in-bulk points. Iron ore moved east from the great mines of the Lake Superior ore field to the well-established iron- and steel-producing centres in Ontario and the eastern United States. Also, several of the ore-shipping ports developed iron and steel industries and Duluth was the chief producer. Consequently, not only was an eastward flow of iron established, but also a return flow of coal and coke from the eastern coalfields. Sault Ste. Marie intersected these flows of the major raw materials for heavy industry.

Another element was the potential for the generation of electricity, particularly hydraulic electricity. The late 19th century was as euphoric over the promise of the applications of electricity as the early 19th century had been about steam and the Sault rapids attracted particular attention.

Given raw materials and energy and knowing that labour can always be attracted to an expanding economy, only the infusion of capital was required to realize the economic development of the Sault. Francis H. Clergue provided the capital and the essential "magnetism" and "ambition" Sullivan referred to.[5]

In 1888 the newly formed St. Mary's Falls Water Power Company began excavation of a 150-foot-wide canal to divert water power from above the rapids to mill sites several miles below.[6] The venture was abandoned the following year and lay dormant until purchased by Francis H. Clergue, who was visiting the town. Reorganized as "The Lake Superior Power Company" in 1895, the new company acquired the stock held by the town of Sault Ste. Marie in "The Ontario and Sault Ste. Marie Water Light and Power Company," which had taken over from the original power venture.[7] Between 1894 and 1903, Clergue and his associates moved into such diverse activities as a local utility company, a sulfide plant, a pulp mill and an iron and steel works. All these production units were centred on the rapids and the application of their power. They in turn prompted investments in ore mines and the development of transport facilities including railways and a shipping company.[8]

By 1902 Consolidated Lake Superior Corporation, the industrial empire controlled by Clergue, was an integrated economic system approaching the diversity and complexity of a modern conglomerate. Despite the Canadian names of its constituent elements, however, most of its $148 million stock was held by Pennsylvanian investors. Together with several sulfur, gold, iron and nickel mines, some of the main holdings were Lake Superior Power Company; Sault Ste. Marie Pulp & Paper Company; Tagona Water & Light Company; Michigan & Lake Superior Power Company; Algoma Steel Company; Algoma Central and

Hudson Bay Railway; and the Ontario, Hudson Bay and Western Railway. Perhaps because of this diversity and substantial overextension, economic collapse, profound labour problems and government intervention followed the initial decade of expansion and great promise. The various elements of the industrial economy struggled in reorganized forms as the Sault economy reflected the economic fortunes of the rest of Canada during the twenties and thirties. At the onset of World War II the economy flourished and Algoma Steel and Abitibi Pulp and Paper, both progenies of the Clergue empire, were the new economic dominants of the town.

Statistics of the manufacturing production in the town during the half-century of development between 1881 and 1951 period demonstrate the continued increase in industrial activity:[9]

	No. of firms	No. of employees	Payroll in dollars	Value of products
1881	-	315	94 833	299 647
1891	20	101	29 665	107 510
1901	13	665	297 258	738 472
1911	23	501	345 581	1 002 834
1921	-	-	-	-
1931	47	1692	2 757 577	12 672 466
1941	49	5574	9 486 497	44 056 386
1951	53	8008	25 545 411	133 911 798

27 Industrial complex at Sault Ste. Marie. (SSMCC)

In terms of the labour force, the proportion of the population employed in manufacturing increased from 21% in 1871 to 53.4% in 1941.

From the first years of the operation of the canal, therefore, Sault Ste. Marie was an industrial town. As such, it was well served by the transport facilities provided by the canal and the shipping that focussed on the narrow corridor through the Sault. Proximity of transportation facilities and industrial plant did not always lead to a harmonious relationship. Situated as it was in the rapids, the canal reserve was soon bracketed by actual and potential power projects. To the north of St. Mary's Island, the Sault Ste. Marie Light and Power Company (then controlled by the town but later to become Clergue's Lake Superior Power Company) cut its canal above the rapids and became established as the major power producer on the Canadian side. Indeed, in 1893 the company, then virtually a municipal utility, made an abortive bid to supply the lock with the power, arguing:

> The Ontario and Sault Ste. Marie Water, Light and Power Co. expect to complete their water power canal this autumn, and be in a position to furnish light or power to any desiring the same. The ship canal, now approaching completion, will we presume require to be lighted by electricity next summer, and the Board would be pleased to furnish you with an estimate of the amount for which they would supply the ship canal with light, or with power only, in case the dept. of Railways and Canals wish to put in its own plant for light....The town has undertaken a serious responsibility in the construction of the Power Canal and it is a matter of deep concern to them that such a use may be made of it as will bring some prompt return for their outlay....The Board of Directors respectfully submit their claim for first consideration in the matter of supplying the Ship Canal with Light or Power.[10]

The government, or at least the engineers of the Department of Railways and Canals, were firmly committed to a completely self-contained system and not even the prospect of helping to assist an ailing municipality could lure them away from this intent.

Then again, to the south of St. Mary's Island, several parties were interested in power developments. In April 1889 Ryan was given the use of island No. 10 for construction purposes.[11] In 1892 the contractors applied for islands 1-10, arguing they had invested some $50 000 on the machinery, plant and outfit there. Although denied ownership, they were allowed to continue to use the group on the same terms as island No. 1. Ryan repeated his request in 1894 and this time he was allowed to buy islands 1-8, and those portions of 9 and 10 east of the land owned by the C.P.R. The price was $5 per acre, or a mere $40 for the entire group. Ryan paid the $40 and was issued a patent on May 7, 1895. The patent was engrossed but not signed by Lieutenant-Governor Sir George Airey Kirkpatrick; it was not registered in the office of the provincial registrar. "Things remained in status quo until 1896," noted an official in the province's Department of Lands, Forests and Mines, "when a petition of the citizens [of Sault Ste. Marie] dated November 9th, 1896, was sent to the Hon. Commissioner of Crown Lands, urging the development of the water power on the rapids of the St. Mary River, south of the Ship

Canal, and urging that the said power be put up at public auction and sold only on condition of its immediate development, in the same manner as the power to the north of the Ship Canal, now owned by the Lake Superior Power Company." Two days before Christmas 1896, the province advertised in the press the sale at auction of islands 1-12. The sale was subject to the condition the purchaser pay Hugh Ryan and Co. for the value of their improvements. A further condition required the purchaser to establish within 1 year a manufacturing industry using at least 200 h.p. The only bid was $6000, which the province considered inadequate. The property was withdrawn. Eventually, however, the Department of Railways and Canals intervened, arguing that increasing

28 Francis Hector Clergue. (Archives of Ontario, S.3843)

traffic flows might soon necessitate an additional lock and that the islands south of the canal should be reserved for canal purposes. Those seeking sites suitable for the generation of extra power at the Sault were, therefore, obliged to look elsewhere.

The next move was Clergue's and it was a dramatic one. South of the islands of concern to the municipality of Sault Ste. Marie lay the largest island in the rapids, Whitefish Island, and by 1902 it appeared to have been acquired by Clergue. Nominally it was for the construction of docks, warehouses and terminals for three railway companies: the Algoma Central and Hudson Bay Railway, the Pacific and Atlantic Railway, and the Ontario, Hudson Bay and Western Railway.[12] Considering the location of the iron and steel plant, these plans appear to make sense as portage lines for incoming coal and outgoing ore and iron products which would bypass the canal. Nonetheless, hydroelectrical power was involved yet again.

The original disposition of the lands was by orders-in-council between January 1900 and January 1902. When knowledge of these developments became known, the first to protest was the Sault Ste. Marie Board of Trade.[13] Armed with the argument that two of the companies "are operating no railway, nor have they any railway under construction," the board expressed its chief concern:

> The matter is one of very great importance to the town of Sault Ste. Marie, and I have no doubt at all but that strong representations will be made to you by the official representative of the town in this matter. On the north side of the Island there is a water power in which the town is interested and which is known as the Ryan-Haney Water Power. Beyond the water privileges controlled by the Lake Superior Corporation this is the only independent power available at the present time, and it is of the utmost importance to the town that its interests therein should be conserved and protected in the strongest possible manner....[14]

Predictably, the Lake Superior Power Company wanted no competition. Development of the then dormant Ryan-Haney water power, argued the private sector hydro company, would "delay and hinder the future necessary improvements in the interests of navigation and commerce." Reserving "these lands for public uses, in the interests of navigation," continued the firm, would not "prevent the utilization of the full amount of water available...as the largest development can be answered by diversion and not by development in the rapids."[15] The Lake Superior Power Company's water supply was acquired by diversion above the rapids.

On June 21, 1906, *The Sault Daily Star* reported that the Royal Commission on Transportation had concluded "while the locks at Sault Ste. Marie are, at the present time, capable of providing for all vessels at that point, still the requirements of the future may demand an additional canal and the commission recommend that the government obtain at an early date all such land as may be reasonably required, in case the construction of an additional canal be at any time decided upon." There was particular concern over the fortunes of Whitefish Island, which had been patented by the three Clergue railway ventures.

Although all the land there had thus been severed from the public domain, little had been done in the way of railroad development other than the possible construction of one spur line by the Algoma Central. The concern for the reservation of these lands for future canal development was paramount, however, and in 1913 Whitefish Island was expropriated. Despite a flurry of canal fever during 1913, again in 1941 and once more in the 1950s during the actual development of the St. Lawrence Seaway, there were to be no further canal developments at the Canadian side of the Sault rapids. The expanded reserve of Crown lands remained as part of the establishment there and became integrated into the public land use of the canal grounds.

Social Links

The role of the Canadian ship canal in the life of the local community was not merely economic. Its management was integrated into the social fabric of the society. More than anyone else, J.W. LeBreton Ross promoted this involvement, acting in several capacities within the community of Sault Ste. Marie during his tenure as superintending engineer from 1907 until his retirement in 1936. As president of the local historical society, Ross was a frequent speaker at public occasions and an established authority on the history of the area. He also served as president of the Sault Ste. Marie Horticultural Society and was active in its campaign to "Beautify the city and increase the interest in flower and tree planting."[16] More than this, however, he was interested in the well-being and advancement of the community, perhaps best demonstrated by his concern for the full and effective use of the canal grounds by the people of the Sault.

Certainly, as superintending engineer of the canal, he set a good example for his fellow members of the horticultural society by developing the gardens and grounds at the lock site. This also benefitted the local community as the lock and adjacent lands during the first half century of canal operation were very much part of the public amenities of the municipality. Superintendents Boyd and Ross had been energetic advocates of the need for the allocation of funds for the improvement and beautification of the grounds at the canal. They received support from General Manager Charles Melville Hays of the Grand Trunk Railway when he visited the Sault and spoke of the "fine chance there was to make a park at the canal" and said he "would be glad at any time to lend a hand in convincing the government that some money should be spent in beautifying the canal grounds."[17] The Sault Ste. Marie Board of Park Management also exerted pressure and in June 1911 the minister of Railways and Canals informed the group he had decided that "steps will be taken to improve the Gov't property about the Soo canal, and that the sum of $6,000 has been set apart for this purpose."[18]

Not long after the work had begun, events interfered with the general access by the public. During World War I, national security took precedence over public services and the recreational use of the site was temporarily stopped. At end of the war, one of the first signs of the onset of peace and normal routine was the removal of the "Sault Canal Guard." The community's response was enthusiastic:

> For four years, the ship canal which was one of the first sights which visitors to the Sault looked for, has been closed from public view and Whitefish Island, which was a beauty spot on the city's waterfront before the war, has been out of bounds of the public as a recreation ground. The privilege of returning to the old summer resort next year will give the citizens considerable satisfaction, as the familiar demand for passports will no longer be heard.[19]

As with any industrial towns that grew rapidly and beyond their original plans, the Sault lacked public parks. Therefore the amenity at the ship canal became more important as a recreational site for the community.

This relationship between the populace and the canal establishment was not without occasional friction. In 1921 six men were arrested and charged with "entering an engine room in a building at the locks and using it as a dressing room to accommodate them in diving off the canal wall."[20] Generally, however, despite the busy activity at the lock, the staff appeared to be able to accommodate the demands placed on it by the general public. Thus, in June 1922 about 50 "young people" of the "Epworth League and Mission Circle" spent an enjoyable evening at the "canal park" with a picnic "lain in true picnic style on the grass"; also, "The members of the party explored Whitefish Island. It was thought that the canal is an ideal place and a very happy time was enjoyed."[21]

These halcyon times were threatened in the thirties and the deterioration of the relationship between the public and the canal staff was such that Ross requested police supervision of the site. The citizenry complained of "Immorality, Petting, Nude Bathing, and Property Destruction," and the *Daily Star* (Apr. 29, 1933) trumpeted abroad that "Whitefish Island And Canal Grounds Should Be Cleared Up Say Citizens." The same article reported Ross's comments on the whole affair: "We want the people of the Sault to enjoy the canal grounds but there is very little encouragement for us to go on and improve the grounds further if they are to be damaged and destroyed by thoughtless people who visit and use them." After two decades of developing his grounds into a local beauty spot, he must have been very disappointed.

Even so, attention continued to be focussed on the government lands at the canal reserve for development as a public park. The focus shifted somewhat to the adjoining Whitefish Island, a relatively neglected property but one long used as an informal recreational retreat by the local community. The demand for work in a depressed economy also fitted into the proposal for the development of an island park; the Sault Ste. Marie City Council moved that "we ask Mr. J.E. Simpson, our present Member at Ottawa, to take up the matter of securing permission and help in clearing up Whitefish Island, to be used as a park, and as a means of providing work for those on relief."[22] Ross reported to Ottawa that "The Island is thickly covered with brush which could be cleared up to advantage to enable the public to move around, and this is about the only work which could be done, as any attempt to make a park of it would require permanent labour to keep it in order." Because there was no intention to charge the expense to government, Ross recommended "the city be allowed to place men on the island under my supervision."[23] Ottawa agreed to this. A. Dubuc, chief engineer of Railways and Canals, replied to Ross: "authority is granted for the cleaning up of Whitefish Island by the Sault Ste. Marie City Council on condition that the work is done under your supervision and creates no obligation on the department to maintain the area as a park."[24]

To the last years of his superintendency at the Sault, though somewhat considerably disillusioned by the lack of appreciation, Ross strove to maintain the park as a beautiful site as well as an efficient element of the Canadian navigation system. The latter role indirectly benefitted the Canadian society at large; the provision of a park and the protection of the public's right of access directly benefitted the people of the Sault.

Protecting the Lock

Never was the national importance of the Canadian lock more recognized than in wartime. Perhaps it was the recollection of past incursions from ill-disposed persons among the neighbours to the south or else the frequently advocated vulnerability of the lock to sabotage. Whatever the reason, two world wars prompted extraordinary defensive measures on the part of the government.

After the declaration of war in 1914, Superintendent Ross recognized the need for vigilance and for appropriate procedures to protect his facility:

> Upon the outbreak of war on August 4 last, a guard was placed on the canal property to protect it from damage. Also, at the request of the military authorities, additional linemen were placed on duty to handle the lines, so that it would be unnecessary for vessels to put their own linemen ashore.[25]

When the guards were in place, the canal grounds took on a new interest for the public in the first few years of the war. Indeed, there was even a festive air as the 51st "Soo" Rifles put on a public "rifle shoot" at the "miniature range" on the canal grounds, followed by a demonstration of "the Force and Effect of Bullet and Steel,"

> showing the method employed by the British army in storming an opposing and well protected position. The company formed at the east end of the lock, and stormed the heights at the western end, which were protected by the men on duty along the canal wall.... With characteristic British cheers the company, to the strains of the bugle giving the charge, stormed the heights, with fixed bayonets, drove the defenders pell mell from their position, which was presumably fortified.[26]

By December 140 officers and men were stationed at the canal, the officers being quartered in the "main canal building"; other ranks were billeted in the "old cement shed" on the north pier which had been converted into a barracks, mess and kitchen; others present included 13

29 Troops beside the administrative building, World War I. (SSMCC)

interned "enemy aliens" who were "Made to Work a little to work up an Appetite"; they were accommodated in a box car on the Algoma Central Railway spur line.[27] Apart from manning the three guard posts to protect the canal and lock, the military also trained units for duty overseas and it was reported that "Regular Daily Drill is Held and Men Soon Become Proficient." But it was not all spit and polish -- the same report referred to a military hockey team at the canal and skating on ponds nearby. A more formal social event took place in December 1915 when the British Patriotic League served an oyster supper to the off-duty men of the 51st Regiment at their barracks at the ship canal.

As the "Great War" continued into 1916, pressures for manpower increased and there was a move to assign local "volunteers" to the 51st to free others presently on guard duty at the canal for service overseas:

> It was pointed out at the office today that another twenty men could be released from the guard at the canal for overseas service if men could be secured to take their places and it is the desire of those in charge to get recruits for guard work. The pay of the men on duty at the ship canal is the same as that of men going overseas, $1.10 per day.[28]

The recruitment continued into 1917, the populace being assured that "those who now enlist in the 51st. are not at all likely to be called on to do duty at the canal unless something very extraordinary happens, in which case the Militia Act would certainly be brought into operation and everybody would be summoned."[29] At a public meeting a few nights later, the Rev. H.S. Pritchard alluded further to such "extraordinary" events, querying "what would happen if a band of Germans from across the border, unorganized it may be, should attack our canal or come and make war on our steel plant...? We need to protect them."[30]

Others concerned with the security of the lock reminded Ross of events he would have liked to forget. Mr. C.H. Clark from New York enquired whether "the protection against vessels striking the gates of the canal lock — before entering are the same now as about 7 years ago — if so it appears to me vessels could easily strike the gates as they did a few years ago and do considerable damage — especially during this war."[31] Ross's response was brief, to the point and uncommunicative: "it would not be proper for me to give out any information during war time."[32] When the United States entered the war in 1917, stringent American security interfered with Ross's routine, and he complained to L.C. Sabin, superintendent of the St. Mary's Falls Canal: "I have been in the habit of walking across the bridge, when visiting your canal, but now find that you have a guard at your end of the bridge as well as the guard at our end, so that I am not able to cross."[33] Canadian-American harmony was maintained when Major Wilson, commander of the garrison protecting the U.S. canal, issued Ross a pass. Considering such thorough security, it is no wonder that all the canals at the Sault survived World War I without being damaged.

The threat of possible sabotage or direct aerial attack was taken even more seriously during World War II. The first scare came some months before the actual declaration of war when in April 1939, 136 sticks of dynamite were discovered in a shed on the canal property. The immediate interpretation was that it was stored there for use in crippling the city's power and water service or the Great Lakes shipping movements which were "of vital importance to the Empire in time of war." The *Daily Star* trumpeted (Apr. 22, 1939), "Government Should Act to Protect Canals." By August the international situation had deteriorated to such an extent that the approaches to the ship canal had been barricaded and several vulnerable points were being guarded by soldiers of the Sault Ste. Marie and Sudbury machine gun regiment. The *Daily Star* (Aug. 26, 1939) warned the public that "Persons approaching the guard station and failing to take warning will do so at their own risk." The situation had eased somewhat by November and the soldiers were replaced by 19 ex-servicemen who took over the guard duties.

The Canadians might have relaxed somewhat, but their American neighbours to the south were apprehensive of the possibility of attack from a new direction. Lt.-Col. Harold of Michigan reported that measures were being taken to protect the Sault locks from "Axis bombing attack." His concern was that it was "not beyond the ingenuity of Hitler's military schemers to carry bomber planes in sections across the Atlantic by submarines and slip them through the slim cordons of the naval patrols into the Hudson Bay area" where they could be assembled on some lonely island and made ready to attack the canal. Saboteurs could mount less dramatic but equally effective attacks and Judge J.H. MacDonald lectured members of the Ontario Police Association on their responsibilities in wartime; he made special reference to the ship canal:

> The most vital and vulnerable point of North American commerce is right here in Sault Ste. Marie. To put the locks out of

commission would mean as much as the loss of a battleship or a military campaign. Your special responsibility is to guard against sabotage and disloyal activity.[34]

In any case, there was no attack and the locks functioned without interruption throughout the war. The only shots heard were fired when a machine gun was discharged accidentally and an unfortunate private of the

30   Guarding the lock, World War II. (SSMCC)

31   Defence against air-raids, World War II. (SSMCC)

131st Infantry Battalion was killed.[35] Gradually, however, the wartime fears diminished. One of the restrictions concerned the Great Lakes passenger traffic, no travellers being allowed west of the Sault. On May 21, 1943, the *Daily Star* announced that passengers were to be permitted to continue their journey through the Sault locks to the lakehead ports and return. The locks were returning to their normal routine.

Continuity and Heritage

Apart from the economic contribution, the Sault ship canal has always been very interesting to the local populace and visitors and has also been accepted in the Canadian popular awareness as a significant location. When first constructed, the electrically operated and "longest lock in the world" was in the forefront of national attention. The transcontinental railway had captured Canada's imagination two decades earlier, but the "Soo" was another project that lived up to the expectations of the young nation's pride and self confidence.

Accounts of its construction and pictures of its facilities figured in the popular magazines and professional journals of the day. The lock was even featured on the nation's currency, the 1900 Dominion of Canada $4 bill featuring the Countess and Earl of Minto going through the "Soo" canal, albeit the American one. The error was corrected in 1902 on another Dominion of Canada $4 bill.[36] Perhaps the greatest publicity received by Sault Ste. Marie, its setting and people was that afforded by the 1922 film of Sullivan's novel, *The Rapids*. Produced by Ernest Shipman of New York and starring Mary Astor, 100 prints were made for

32 Advertisement for the film *The Rapids*. (SSMCC)

33 The Sault Ste. Marie ship canal. (SSMCC)

circulation in the U.S., 25 for Canada and 50 for "the rest of the world."[37] This film retold and popularized the story of the development of the isolated trading community into a dynamic industrial complex.

The quarter-century after the opening of the ship canal was heady for Sault Ste. Marie. The buoyant optimism of the period was symbolized by the movie version of *The Rapids*. After the 1920s, all of this changed as the Sault realized that it was not destined for metropolitan greatness. The canal failed to realize the potential dreamed of by its designers and builders, and became a minor cog in the Canadian-American transport system on the St. Mary's River. Nonetheless, the Canadian Sault canal, which was a brilliant achievement of innovative engineering and technology, was part of a vision of national independence and greatness and retains its place in our popular imagination.

ENDNOTES

I The First Contact: The "Bawatig" and The Fur Companies

1. This chapter has benefitted much from Graham A. MacDonald, "The Saulteur-Ojibwa Fishery at Sault Ste. Marie, 1640-1920," MA thesis, University of Waterloo, 1978.
2. Ibid.; Bishop Baraga, A Dictionary of the Otchipwe Language (Minneapolis: Ross and Haines, 1973); W. Vernon Kinietz, The Indians of the Western Great Lakes, 1615-1760 (Ann Arbor: University of Michigan Press, 1940), pp. 317-18; for Dablon's reference see R.G. Thwaites, ed., Jesuit Relations and Allied Documents (Cleveland: Burrows, 1896), Vol. 54, p. 133.
3. Peter Grant, "The Saulteux Indians about 1804." In Les Bourgeois de la Compagnie du Nord Ouest, ed. R.L. Masson (New York: Antiquarian Press, 1960), Vol. 2, p. 330.
4. Duncan Cameron, "A Sketch of the Customs, Manners and Way of Living of the Natives in the Barren Country about Nipigon." In R.L. Masson, op. cit., Vol. 2, pp. 297-98.
5. R.G. Thwaites, op. cit., Vol. 23, pp. 223-27.
6. R.G. Thwaites, op. cit., Vol. 54, pp. 129-31.
7. "Hennepin's Narrative from La Louisiane," Illinois Historical Collection, 1903, p. 52.
8. Alexander Henry, Travels and Adventures in Canada and the Indian Territories, 1760-1776 (Edmonton: Hurtig, 1969), p. 62.
9. Grant, op. cit., p. 330.
10. John Johnston, "An Account of Lake Superior." In R.L. Masson, op. cit., Vol. 2, p. 147.
11. John Askin Jr. to John Askin Sr., Sept. 8, 1807, The John Askin Papers, M. Quaife, ed. (Detroit: Detroit Public Library, 1928), Vol. 2, p. 574.
12. U.S. Commission of Fish and Fisheries, Report of the Commission for 1887 (Washington: Government Printing Office, 1891), Part 15, p. 68.
13. Ibid., p. 36.
14. J. Carver, Travels Through North America in the Years 1766, 1767, 1768 (London: 1778), p. 142.
15. Anna Jameson, Winter Studies and Summer Rambles in Canada (Toronto: McClelland and Stewart, 1965), p. 130.
16. George Catlin, Letters and Notes on the Manners, Customs and Conditions of the North American Indians, 1832-1839 (New York: Dover, 1973), Vol. 2, p. 162.
17. W.C. Bryant, "Sketches of Travel." In Prose Writings of W.C. Bryant, ed. P. Godwin (New York: Russell and Russell, 1964), Vol. 2, p. 69.
18. U.S. Commission, op. cit., p. 68.
19. Canada. Public Archives. Manuscript Division (hereafter cited as PAC), RG10, Vol. 2824, file 168, 291, Hamilton to Sifton, Feb. 25, 1899.
20. Ibid., Vol. 213, Hamilton to L.K. Jones, Secretary, Department of Railways and Canals, Dec. 24, 1901.

21 Ibid., Vol. 2824, file 169,291, Abbott to Indian Affairs, Dec. 13, 1898.
22 Ibid., Peter Kahgayosh to Indian Affairs, Sept. 18, 1902.
23 L.P. Kellog, ed., Early Narratives of the North West, 1634-1699 (New York: Scribner, 1917), p. 217f.
24 See Paul C. Phillips, The Fur Trade (Norman: University of Oklahoma Press, 1961).
25 Thor Conway, The Point Louise Site: Report of The Regional Archaeologist (Sault Ste. Marie: Ontario Ministry of Culture and Recreation, 1975).
26 F.C. Bald, The Seigniory at Sault Ste. Marie (Sault Ste. Marie: 1937); H.P. Bews, The French and British in the Old North West, (Detroit: Wayne State University, 1964), p. 32.
27 W.R. Riddell, "The Last Official Report of the French Posts in the Northern Part of North America." Ontario Historical Society Papers and Records, Vol. 28, 1932, p. 134.
28 F.J. Turner, The Character and Influence of the Indian Trade in Wisconsin: A Study of the Trading Post as an Institution (1891) (New York: Benjamin Franklin, 1970), pp. 43-44.
29 G.C. Davidson, The North West Company (New York: Russell and Russell, 1967), pp. 9-10.
30 H.A. Innis, The Fur Trade, revised by S.D. Clarke and W.T. Easterbrook (Toronto: University of Toronto Press, 1964), p. 183.
31 PAC, MG11, CO42, Vol. 70, pp. 65-70, G. Mann to Lord Dorchester, Dec. 6, 1788 (emphasis his).
32 Province of Ontario. Archives of Ontario (hereafter cited as PAO), Russell Papers, MS75, Reel No. 5, Transfer of Lands by the Natives at Sault Ste. Marie to Simon McTavish, Joseph Frobisher, John Gregory, Ian McGillivary and Alexander Mackenzie, Aug. 10, 1798.
33 PAC, RG8, C Series, Vol. 363, pp. 33-34, Phyn. Inglis and Co. to the Duke of Portland, Dec. 6, 1799.
34 Ibid., March 13, Portland to Hunter.
35 J.W. Ross, "Transportation in the Early Days," Sault Ste. Marie Historical Society Museum, Annual Reports (1920-21), Vol. 1.
36 PAC, RG81, Vol. 382, pp. 215-18, "Report on the Establishment formed by the North Shore contiguous to the Falls of St. Mary by the North West Company Trading to the Indian Country," by Capt. R. Bruyères, R.R., 1802.
37 PAC, RG8, C Series, Vol. 363, pp. 38-41, X.Y. Company to the Governor, Dec. 23, 1803.
38 U.S. National Archives, RG107, Registered Correspondence to the Secretary of War, Microcopy 221, Roll No. 60, Holmes to Lt.-Col. G. Crogham, n.d.
39 PAC, RG8, C Series, Vol. 363, pp. 169-74, Dalhousie to Wellington, April 23, 1824.
40 Ibid., p. 157, H. Darling to Thomas Thain, Agent for the Hudson Bay Company, March 31, 1824.
41 Ibid., p. 156, Ground Plan of the Buildings at St. Mary's; also, pp. 201-202, Capt. Gaff, 76th Regiment to W.A. Thompson, Sept. 6, 1824.
42 Ibid., Dalhousie to Wellington.

43 Hudson's Bay Company Records (hereafter cited as HBC), Sault Ste. Marie Journals, B194/a/1, A Journal of Transactions and Occurrences at Sault St. Mary's commencing Sept. 1, 1824.
44 Ibid., B194/e/1, Sault Ste. Marie Annual Report, 1825.
45 Ibid., B194/e/3, Sault Ste. Marie Annual Report, 1827.
46 Ibid., B194/e/4, Sault Ste. Marie Annual Report, 1828.
47 Ibid., B194/e/5, Sault Ste. Marie Annual Report, 1829.
48 Ibid., B194/a/5, Sault Ste. Marie Correspondence, Simpson to Bethune, July 21, 1829.
49 Ibid., B194/a/1-9, Sault Ste. Marie Journals.
50 Ibid., B194/c/1, Letter from J. Black, Fort Garry, Dec. 19, 1853.
51 Ibid.. B194/b/4, Bethune to Haldane, June 7, 1826.
52 Ibid., B194/a/3, Sault Ste. Marie Journal, 1827-28.
53 Ibid., B194/a/9, Sault Ste. Marie Post Journal, 1835/36.
54 Ibid., B194/b/10, Simpson to Nourie, Sept. 20, 1835.
55 Ibid., B194/b/12, Nourie to Simpson, Sept. 16, 1837.
56 Ibid., B194/b/13, Nourie to Simpson, June 21, 1838.
57 Ibid., Simpson to Nourie, Aug. 11, 1838.
58 Ibid., B194/b/13, Sept. 1, 1835, Nourie to Simpson.
59 Ibid., B194/b/12, June 15, Aug. 8, 1837, Nourie to Leith.
60 Ibid., Aug. 10, 1837, Nourie to Simpson.
61 PAO, W. McTavish Letterbook (HBC), General Box 7, Item 7, Nov. 1848, McTavish to James Hargraves.
62 Ibid., W. Simpson to Hopkins, March 29, 1865.
63 Ibid., B134/c/102, E.B. Barron to E.M. Hopkins, deputy governor general, Hudson's Bay Company, Montreal, May 28, 1866.

II Bypassing the Rapids: The Case for the Canals at the Sault

1 PAO, Department of Crown Lands, Crown Lands papers, Ontario Department of Lands and Forests, Commissioner of Crown Lands Papineau to Alexander Vidal, Dec. 4, 1845.
2 Ibid., Papineau to Vidal, June 3, 1856.
3 Canada. Census. 1861.
4 Ibid., 1871, 1881.
5 H.G. Kingston, Western Adventures or a Pleasure Tour in the Canadas (London, 1856), p. 218.
6 Ibid., p. 195.
7 PAC, HBC, B194/c/1, H. Mackenzie to J. Hargreaves, April 27, 1853.
8 Kingston, op. cit., p. 195.
9 R. Alan Douglas, ed., John Prince: A Collection of Documents (Toronto: The Champlain Society, University of Toronto Press, 1980), p. 128.
10 Ibid., p. 200.
11 PAO, W. McTavish Letterbook (HBC), General Box 7, Item 7, William McTavish to George McTavish, Feb. 6, 1849.
12 F.C. Bald, "The Story of the Sault," The Beaver, p. 51.
13 Canada (Province). Department of Public Works. "Keefer Report," Journals of the Legislative Assembly, 1852-53, Vol. 9, No. 3 (1853), Appendix O: "Sault Ste. Marie Canal," Aug. 19, 1852.

14 <u>Niagara Chronicle</u>, Jan. 2, 1853.
15 Ibid.
16 For information on the construction of the American canal, see I. Neu, "The Building of the Sault Canal." <u>Mississippi Historical Review</u>, Vol. 40 (June 1953), pp. 25-46; F.C. Bald, "The story of the Sault," <u>The Beaver</u>, 286 (Autumn 1955), pp. 48-52; Graham A. McDonald, "The Saulteur-Ojibwa Fishery at Sault Ste. Marie, 1640-1920." MA thesis, University of Waterloo, 1978.
17 Kingston, op. cit., p. 200.
18 <u>Report to a Committee on the Country North of Lake Superior</u>, Toronto <u>Daily Leader</u>, Sept. 27, 1853, in Douglas, op. cit., pp. 122-23.
19 Canada (Province). Dept. Public Works. "Keefer Report," op. cit.
20 <u>Huron Signal</u>, Nov. 4, 1852.
21 <u>Owen Sound Comet</u>, Sept. 24, 1852.
22 W.J. Rattray, <u>The Scot in British North America</u>, (Toronto: Maclear and Co., [1880-84]), Vol. 4, p. 1182.
23 PAC, MG24I8, Vol. 40: notes on "a meeting of the Divisional Directors of the Victoria Mining Company," Aug. 5, 1856.
24 Ibid., HBC, B194/b/15, McTavish to Simpson, Sept. 14, 1849.
25 For these initiatives, see Legislative Assembly Journal, 1851, pp. 43, 182-83, 212; ibid., 1853, p. 901; Gerald E. Boyce "Canadian Interest in the Northwest, 1856-1860." MA thesis, University of Manitoba, 1960; Donald Swainson, "The North-West Transportation Company: Personnel and Attitudes." <u>Historic and Scientific Society of Manitoba, Transactions</u>. Series III, No. 26 (1969-70), passim; Dennis Carter-Edwards, "The Sault Ste. Marie Canal," Research Bulletin No. 119, Parks Canada, Cornwall, 1980.
26 Canada. Census, 1861-62, 1871-72, 1881-82.
27 <u>Globe</u> (Toronto), March 24, 1847.
28 Ibid., June 14, 1848.
29 Ibid., Dec. 4, 1856.
30 Province of Canada, <u>Statutes</u>, 1858, pp. 635ff.
31 PAC, HBC, B194/b/12, Nourie to Simpson, Feb. 9, 1838.
32 Ibid., Nourie to Simpson, March 1, 1838.
33 Ibid., B194/b/13, Nourie to James Leith, May 17, 1838.
34 Ibid., Nourie to Col. Foster, July 20, 1838.
35 Ibid., Nourie to Simpson, Sept. 14, 1838.
36 Ibid., H.B.C., B194/b/16, Oct. 12, 1838.
37 PAC, WO55, Jan. 25, 1841, Vol. 876, pp. 19-49ff., "Memorandum on the fortifications at Kingston and on the Defence of the Frontier."
38 Ibid., B134/c/102, W. Simpson to E.M. Hopkins, March 29, 1865.
39 Ibid., June 7, 1866.
40 <u>Globe</u> (Toronto), May 13, 1870; see also May 5 and 19, 1870.
41 <u>Constitutional</u> (St. Catharines), May 19, 1870.
42 PAC, HBC, IM375: Simpson to Mactavish, Oct. 16, 1870.

III Construction of the Canadian Canal

1 Canada. Parliament, <u>Sessional Papers</u> (Ottawa: Queen's Printer)

(hereafter cited as Sessional Papers), 1871, No. 5, pp. 2-3, Memorandum by Hector L. Langevin, July 4, 1870.
2 PAC, RG11, Vol. 43, subject 3: Sault Ste. Marie Canal.
3 George Fletcher Henderson, Federal Royal Commissions in Canada 1867-1966: A Checklist (Toronto, 1967), p. 4.
4 Sessional Papers, 1871, No. 54, pp. 4-5, The Commission to the Members of the Royal Commission, Nov. 16, 1870.
5 Ibid., p. 108.
6 Ibid., p. 39.
7 Ibid., p. 41.
8 Thomas F. McIlwraith, "George Laidlaw," Dictionary of Canadian Biography, Vol. 11, pp. 481-83.
9 Sessional Papers, 1871, No. 54, Supplementary Returns, p. 2, G. Laidlaw to J.C. Arkins, Feb. 28, 1871, in Canada.
10 Ibid., pp. 2-3.
11 W.L. Morton, Manitoba — A History (Toronto, 1957), p. 182.
12 Patrick Burrage, 125 Years of Progress (Winnipeg: James Richardson and Sons Ltd., 1982), n.p.
13 Canada. Parliament, Debates of the House of Commons (Ottawa: Queen's Printer) (hereafter cited as Debates), 1887, II, 830, June 7, 1887.
14 Sessional Papers, No. 20, 1901, Sault Ste. Marie Canal, Superintendent's Annual Report, Aug. 6, 1900.
15 Debates, 1887, p. 831.
16 Ibid., Sept. 9, 1891.
17 Ibid., 1888, II, p. 1442, May 15, 1888.
18 Ibid.
19 Parks Canada, Cornwall Office, Sault Ste. Marie Records (hereafter cited as PCC), File H C4250/S32-1, Vol. 1, C, Maintenance and Repairs, Canal, Page to Pope, April 26, 1888.
20 Sessional Papers, 1889, No. 10. Here "hydraulic" refers to machinery operated directly by water power rather than hydroelectric power. This is discussed more fully in the following section.
21 PCC, File C4250/S32-1, Vol. 4 C, Maintenance and Repairs, Canal.
22 Ibid., Vol. 6C.
23 Ibid.
24 PCC, File C4250/S32-1, Vol. 6C, Maintenance and Repairs, Canal.
25 PAC, RG43, Vol. 1771, file 4860. There were also some incomplete tenders.
26 Ibid., File 4860; Order-in-council, Nov. 12, 1888.
27 PAC, RG43, Vol. 1698, File 4860.
28 Ibid., Van Horne to Macdonald, Aug. 17, 1890.
29 Ibid., Thompson to Bradley, Sept. 18, 1890.
30 PAC, RG43, Vol. 1698, File 4860. Van Horne to Macdonald, April 3, 1891.
31 Ibid. Copy of a resolution of marine section of the Board of Trade of the City of Toronto, March 21, 1891.
32 PAC, RG43, Vol. 1698, file 4860. Trudeau to Macdonald, May 14, 1891.
33 Ibid., Order-in-council, May 21, 1891.

34   Ibid., Department of Railways and Canals to H. Ryan, May 29, 1891.
35   Ibid., Order-in-council, June 4, 1891.
36   Ibid., Trudeau to Bowell, Dec. 17, 1891.
37   Ibid.
38   Debates, May 6, 1892.
39   For the nature of his claims, see PCC, File C4250/S32-1, Vol. 4, C, Maintenance and Repairs, Canal; Schreiber to Balderson, March 3, 1896, and PAC, RG43, Vol. 1698, file 4860, Ryan to Blair, March 9, 1897.
40   Parks Canada, Sault Ste. Marie, Report of the Arbitration Hearing into the Construction of the Sault Ste. Marie Canal (Ottawa: 1899).
41   Sessional Papers, No. 10, 1891.
42   Ibid., 1890, No. 5.
43   Ibid.
44   PAC, RG43, B1, Vol. 421, Series F, Vol. 4, No. 123588, from John Page, March 21, 1889.
45   Ibid., No. 120420, from Interior Department, July 10, 1888.
46   Ibid., No. 75691, to John Page, July 25, 1889.
47   Ibid., No. 120570, from John Page.
48   PCC, DOE, File No. C8616/S32-6, Vol. 1, Page to Secretary of Railways and Canals, June 1, 1889.
49   Ibid., Crown Lands Department, Toronto, to Page, June 5, 1889.
50   PCC, DOE, File No. C-8616/S32-3, Vol. 1, Order-in-council, Feb. 2, 1892, Deed transferred from the Department of the Interior to the Department of Railways and Canals.
51   Sessional Papers, 1891, No. 10.
52   PAC, RG43, Vol. 1697, File 4860, Ryan to Page, Dec. 20, 1889.
53   Ibid., Ryan to Page, May 23, 1890.
54   Ibid., Ryan to minister of Railways and Canals, Feb. 22, 1890, Valuation of Property Situated in the Town of Brockville, March 14, 1890; Order-in-council, April 1, 1890.
55   Sessional Papers, 1891, No. 10.
56   PCC, File C-4250/S32-1, Vol. 2, Allan and Fleming to Page, Dec. 23, 1889.
57   Ibid.
58   Ibid., McDougall to Bradley, Feb. 7, 1890.
59   Ibid., Thompson to Page, July 19, 1889.
60   PAC, RG43, Vol. 864, Order-in-council, Aug. 7, 1890.
61   Sessional Papers, 1892, No. 10.
62   Ibid., No. 9.
63   Ibid., p. 140.
64   PAC, RG43, Vol. 1698, File 4860, Ryan to Bradley, Oct. 17, 1890.
65   The Sault Daily Star (Sault Ste. Marie), Dec. 24, 1952, "Builder of Sault Canal Stern, Fair with Labor."
66   Parks Canada. Sault Ste. Marie, John D. Bouchard, History of Sault Ste. Marie Canal (1967). Parks Canada, Canal Archives, Sault Ste. Marie.
67   Sessional Papers, 1892, No. 9, 141.
68   PCC, File C4250/S32-1, Vol. 2, Allan and Fleming to Trudeau, Dec. 26, 1890.

69 Sessional Papers, 1893, No. 9, 142.
70 Ibid. and Sessional Papers, 1892, No. 10.
71 Ibid., p. 143.
72 PCC, File C4250/S32-1, Vol. 3, Progress Report re. Allan and Fleming, Dec. 4, 1891; Allan and Fleming to Trudeau, Dec. 11, 1891.
73 Ibid., Thompson to Secretary, Department of Railways and Canals, Dec. 17, 1891.
74 Sessional Papers, 1893, No. 9, 145-46.
75 Ibid.
76 Ibid., 1894, No. 6, 121.
77 Ibid., 122.
78 PAC, RG43, Vol. 1698, File 4860, Sinclair to Department of Railways and Canals, June 21, 1892.
79 Ibid., B1, Vol. 421, file 142,000, from C.P.R., Oct. 22, 1892.
80 Sessional Papers, 1894, No. 10, p. 123.
81 Parks Canada, Cornwall Archives, File C-42501, S32-1, Vol. 4, Allan and Fleming to chief engineer of Canals, Jan. 18, 1893.
82 Ibid., Report of a committee of the honourable the privy council approved by the governor general in Council, Feb. 9, 1893.
83 Sessional Papers, 1894, No. 10, p. 123.
84 Ibid., 1895, No. 10, p. 47.
85 Ibid., p. 122.
86 PAC, RG43, Vol. 1698, File 4860 Ryan to Crawford, Jan. 1894; PCC, C4250/S32-1, Vol. 6, Letter to Balderson, Feb. 7, 1894.
87 Parks Canada, Sault Ste. Marie, Correspondence, J.W. Ross.
88 Sessional Papers, 1896, No. 10, p. 234.
89 Ibid., 1895, No. 10, p. 47.
90 Ibid., p. 122.
91 Ibid., 1896, No. 10, p. 129.
92 Ibid.; see also Parks Canada. Sault Ste. Marie, Superintendent's Annual Report, The Electrical Engineer, Vol. 20, No. 389, Oct. 10, 1895: "Power Transmission — The Canadian Ship Canal at Sault Ste. Marie and its Electrical Operation." (hereafter cited as The Electrical Engineer)
93 Ibid.
94 Sessional Papers, 1896, No. 10.
95 PCC, File C4250/S32-1, Vol. 11, C, Maintenance and Repairs, Canal.
96 Sessional Papers, 1896, No. 10, p. 49.
97 Ibid., p. 12.
98 Ibid., p. 234.
99 Ibid.
100 PAC, RG43, Vol. 1698, File 4860, Shanly to Jones, April 15, 1899.
101 Ibid., Order-in-council, Feb. 12, 1900.
102 PAC, RG43, Box 864, Crowley to Blair, March 26, 1898. This letter is transcribed exactly as it was written by Mrs. Crowley. It is an interesting social document. The term "Goolie" is one often applied to prairie Icelanders and the unclear script may be referring to such people rather than the family "Gorby."

IV Operating the Canal

1. Sessional Papers, 1897, No. 10; Parks Canada. Sault Ste. Marie, Superintendent's Annual Report, 1896. Most of the material for this section relating to the operation of the canal is derived from the annual reports made by the superintendent, engineer and, later, the superintending engineer.
2. Parks Canada. Sault Ste. Marie, The Electrical Engineer.
3. Ibid., Superintendent's Annual Report, 1897.
4. Ibid., 1911.
5. Canada. Department of Transport, Report of General Superintendent of Canals, Sault Ste. Marie Canal, Annual Report 1942-43.
6. Ibid., 1943-44.
7. Canada. Department of Railways and Canals, Annual Report, Sault Ste. Marie, 1896.
8. Ibid.
9. Ibid., 1897.
10. Ibid., 1898.
11. Ibid., 1896.
12. Ibid., 1909; this report provides a summary of lockage statistics, 1895-1908.
13. Parks Canada. Sault Ste. Marie Canal Archives, Operation of the Locks, J.W. Ross, 1913.
14. Ibid., Lockmaster's Journal 1911. Only four of these were found -- details are provided of lockages for four seasons, giving the names of vessels and date and time of arrival at the locks.
15. Canada. Department of Railways and Canals, Annual Report, Sault Ste. Marie, 1900.
16. Parks Canada. Sault Ste. Marie, Canal Archives, Estimates, Appropriations and Expenditures, 1895-1934; see also "Regulations for the Dominion Canals, 1895" (Ottawa: Government Printing Bureau, 1897); also, appended typescript dated May 13, 1905, Duties of Lockmaster.
17. Canada. Department of Railways and Canals, Annual Report, Sault Ste. Marie, 1896.
18. Ibid., 1897.
19. Ibid., 1899; for more details of the physical establishment, see Parks Canada, Sault Ste. Marie, Canal Archives, General Correspondence, 1908-1919.
20. Ibid., 1896.
21. Ibid., 1897.
22. Ibid., 1908.
23. PAC, RG43, B1, Vol. 96, C.R. Howard, Manager C.P. Telegraphs, to C. Schreiber, ministry of Railways and Canals, April 25, 1898.
24. Dates and statistics relating to the operation of the post office at the Sault Ste. Marie site are obtained by cross referencing two sources: Annual Reports of the Postmaster General, 1910-33, and Canada Official Postal Guide (Ottawa: Government Printing Bureau), 1910-33.
25. Operation of Locks, op. cit.
26. Canada. Department of Railways and Canals, Superintendent's Annual Report, Sault Ste. Marie, 1896.

27 Ibid., Engineer's Report, Sault Ste. Marie, 1904.
28 Ibid., Supterintendent's Annual Report, Sault Ste. Marie, 1907.
29 Ibid., 1919.
30 Ibid., 1926.
31 PCC, File C4256/532-10, Vol. 2, Construction, Maintenance and Repairs; Dams, Weirs, and Swing Dam; Specifications for Repairs to be Made to the Steel Moveable Dam, Sault Ste. Marie Canal, Dec. 21, 1910.
32 Canada. Department of Railways and Canals, Superintendent's Annual Report, Sault Ste. Marie, 1901.
33 Ibid., 1912.
34 Ibid., 1909.
35 PAC, RG43, Department of Transport, Canal Branch, Vol. 1717, File No. 7975, Series of Telegrams regarding Perry G. Walker accident.
36 The description of the accident and its aftermath is based on two documents: Department of Railways and Canals, Sault Ste. Marie Superintendent's Annual Report, 1910; Parks Canada, Sault Ste. Marie, Canal Archives, J.W. Ross to W.A. Bowden, chief engineer, regarding accident, June 18, 1909.
37 "The Accident at the Canadian Lock at Sault Ste. Marie," The Engineering Record, Vol. 59, No. 25, p. 790.
38 Parks Canada. Sault Ste. Marie, Canal Archives, manuscript "Thrilling Accident at Sault Ship Canal," June 9, 1909.
39 The Engineering Record, op. cit.
40 Ibid.
41 Ibid., p. 792.
42 Canada. Department of Railways and Canals, Superintendent's Annual Report, Sault Ste. Marie, 1910.
43 Ibid.
44 Canada. Department of Railways and Canals, Superintendent's Annual Report, Sault Ste. Marie, 1899.
45 PAC, RG10, Vol. 2824, file 168,291, February 10, 1906, J. Pedley, deputy superintendent general Indian Affairs, to deputy minister Justice.
46 Canada. Department of Railways and Canals, Engineer's Report, Sault Ste. Marie, 1910.
47 Parks Canada. Sault Ste. Marie, Canal Archives, manuscript "Operation of the Locks," n.d.
48 Canada. Department of Railways and Canals, Superintendent's Report, Sault Ste. Marie, 1903.
49 Ibid., Engineer's Report, 1904.
50 Ibid., Superintendent's Report, 1906.
51 Parks Canada, Sault Ste. Marie, Canal Archives, File 20G, "Expropriation of Whitefish and Other Islands," president, Lake Superior Power Company to Hon. F. Cochrane, minister of Lands, Forests and Mines, April 16, 1909.
52 Canada. Department of Railways and Canals, Engineer's Report, Sault Ste. Marie, 1910.
53 Canada. Department of Railways and Canals, Superintendent's Annual Report, Sault Ste. Marie, 1915.

54 Ibid., 1916.
55 Much of the information regarding the changing technology of shipping is derived from George C. Cuthbertson, Freshwater: A History and a Narrative of the Great Lakes. (Toronto: Macmillan of Canada, 1931); James P. Barry, Ships of the Great Lakes: 300 years of navigation. (Berkeley: Howell-North Books, 1973); Walter Havighurst, The Long Ships Passing: The Story of the Great Lakes. (New York: Macmillan and Co., 1942); Lorenzo Marcolin, "Canadian Pacific Railway Company Steamships Lines — Last of an Era." Inland Seas, Vol. 22 (1966), pp. 3-16.
56 Newsletter, Inland Seas, Vol. 22, No. 1, 1966, p. 154.
57 PCC, File No. C4250/532-1, Vol. 12, CMR-Canal, Data from Miles-Ton Reports of Lake Commerce through Canals at Sault Ste. Marie,...1898.
58 Canada. Department of Railways and Canals, Superintendent's Annual Report, Sault Ste. Marie, 1898.
59 Ibid., 1899.
60 Ibid., 1900.
61 Sessional Papers, No. 20, 7-8 Edward VII, A, 1908.
62 Dana T. Bowen, "The Old Lake Triplets," Inland Seas, Vol. 1, No. 1, 1945, pp. 8-12.
63 The discussion of the traffic through the Sault is derived from Canada, Sessional Papers; Annual Reports of the Department of Railways and Canals, 1896-1925; Canadian Dominion Branch of Statistics, Transportation Branch: Canal Statistics, 1925-46.
64 Parks Canada, Sault Ste. Marie, Canal Archives, Superintendent's Annual Report, April 7, 1923, manuscript.

V Epilogue: The Rapids, The Canal and the Town

1 Alan Sullivan, The Rapids (Toronto: University of Toronto Press, 1972), pp. 15-17.
2 Statutes of Ontario, 50 Vic., 1887, cap. 64, "An Act to incorporate the Town of Sault Ste. Marie," assented to April 23, 1887.
3 Canada. Census, 1861-1951.
4 Ibid.
5 Sullivan, op. cit.; for information on Clergue's career, see "Bibliography" for Donald Eldon and for Margaret van Every.
6 Parks Canada. Sault Ste. Marie, Canal Archives, Terry S. Reynolds, "The Soo Hydro: A Case Study of the Influence of Managerial and Topographical Constraints on Engineering Design," Industrial Archaeology, offprint.
7 Statutes of Ontario, 58 Vic., c.119, 1895.
8 A. Sullivan, op. cit., "Introduction" by Michael Bliss.
9 Canada. Census, 1881-1951.
10 PCC, File C4250/S32-1, Vol. 4, C, Maintenance and Repairs, Canal, Bessingthwaite to Haggert, Aug. 22, 1893.
11 PAC, RG43, Vol. 1718, File 10796, memorandum, Department of Lands, Forest and Mines, Toronto, Dec. 5, 1906.
12 PAC, RG10, Vol. 2824, file 168,291, Feb. 10, 1906, Indian Affairs to minister of Justice.

13 Ibid., June 22, 1906, Resolution of Board of Trade, Sault Ste. Marie.
14 Ibid., Jan. 2, 1906, A.C. Boyce to H.R. Emmerson, minister of Railways and Canals.
15 PCC, File No. C-8616/S32, Vol. 1, Land Acquisition, Sault Ste. Marie.
16 The Sault Daily Star, April 9, 1926.
17 Ibid., May 25, 1911.
18 Ibid., June 29, 1911.
19 Ibid., Nov. 25, 1918.
20 Ibid., July 27, 1921.
21 Ibid., June 20, 1922.
22 PCC, C8616/S3-1, Acquisition, Resolution of Sault Ste. Marie City Council re. Whitefish Island, Aug. 27, 1934.
23 Ibid., J.W. Ross to A. Dubuc, chief engineer, Department of Railways and Canals, Sept. 14, 1934.
24 Ibid., A. Dubuc, to J.W. Ross, Oct. 3, 1934.
25 Canada. Department of Railways and Canals, Annual Superintendent's Report, Sault Ste. Marie, 1915.
26 The Sault Daily Star, Sept. 21, 1914.
27 Ibid., Dec. 12, 1914.
28 Ibid., Aug. 15, 1916.
29 Ibid., Feb. 13, 1917.
30 Ibid., Feb. 20, 1917.
31 Parks Canada, Sault Ste. Marie, Canal Archives, General Correspondence, C.H. Clarke to J.W. Ross, May 16, 1917.
32 Ibid., Ross to Clark, May 21, 1917.
33 Ibid., Ross to L.C. Sabin, April 27, 1917.
34 The Sault Daily Star, Aug. 6, 1941.
35 Ibid., July 11, 1942.
36 The Charlton Standard Catalogue of Canadian Paper Money (Toronto, 1980), p. 39.
37 The Sault Daily Star, Sept. 20, 1922.

GLOSSARY

BOARD FOOT   A unit of measure, hence board measure, equal to a board 1 ft. square and 1 in. thick.

CANAL PRISM   Tapered floor of lock chamber or canal providing the maximum draught for boats.

COFFER DAM   A water-tight enclosure used for obtaining a dry foundation for canal and other construction.

CRIBS   A rectangular frame of logs secured under water to form a pier or dam.

CUBIC FOOT   A unit of measure equal to 1 ft. long, 1 ft. wide and 1 ft. thick.

CULVERTS   A channel of masonry or brick work to carry water under a channel or lock chamber.

GUARD GATES   Supplementary gates to the main canal gates.

HOLLOW QUOINS   A recess in the walls at each end of a canal lock to receive the heel post of a gate.

LINEAR FOOT   A unit of measure in which 1 ft. is equal to 12 in.

LOCK CHAMBER   The specific part of a canal controlled by gates and culverts in which a boat is raised or lowered.

LOCK GATES   "Doors" to the lock chamber which open or close to provide access for boats passing through and which retain water in the chamber while vessels are locking through. When closed the gates present an angular face to the current.

MASONRY   Stone used to construct the canal cut and lock chamber.

MITRE SILL   The sill on which a lock gate rests when closed.

PUDDLE TRENCHES   Trench packed with earth or clay to render it watertight.

BIBLIOGRAPHY

Alcock, J.
"Past and Present Trade Routes to the Canadian North West." Geographical Review, 1920.

Antick, J.
"The Fur Trade in Eastern Canada until 1870." Manuscript Report Series No. 207, Parks Canada, Ottawa, 1976.

Bald, Cleveland
"The French Seigneury at Sault Ste. Marie, Michigan." Sault Ste. Marie Evening News (April 8, 1937).

Bald, F.C.
"The Story of the Sault." The Beaver, No. 286 (Autumn 1955), pp. 48-52.
---. The Seigniory at Sault Ste. Marie. Sault Ste. Marie, 1937.

Ballert, Albert G.
"The Great Lakes Coal Trade: Present and Future." Economic Geography, Vol. 29 (1953).
---. "Commerce of the Sault Canal." Economic Journal of Canada, Vol. 33 (1954), pp. 135-48.
---. "The Soo versus Suez." Canadian Geographic Journal, Vol. 53 (Nov. 1956), pp. 160-67.
---. "Commerce of the Sault Canals." Economic Geography, Vol. 33 (April 1957), pp. 135-48.

Baraga, Bishop
A Dictionary of the Otchipwe Language. Reprint of 1878 ed. Ross and Haines, Minneapolis, 1973.

Barry, James P.
Ships of the Great Lakes: 300 Years of Navigation. Howell-North Books, Berkeley, 1973.

Bews, H.P.
The French and British in the Old North West. Wayne State University, Detroit, 1964.

Bigsby, John J.
The Shoe and the Canoe or Pictures of Travel in the Canadas. London, 1950. 2 vols.

Bowen, Dana T.
"The Old Lake Triplets." Inland Seas, Vol. 1, No. 1 (1945), pp. 8-12.

Boyce, Gerald E.
"Canadian Interest in the Northwest, 1856-1860." MA thesis, University of Manitoba, 1960.

Bryant, W.C.
"Sketches of Travel." In Prose Writings of W.C. Bryant. Ed. P. Godwin. Russell, New York, 1964. Vol. 2.

Burpee, L.J.
"Highways of the Fur Trade." Transactions of the Royal Society of Canada, 1914.

Burrage, Patrick
125 Years of Progress. James Richardson and Sons, Winnipeg, 1982.

Cameron, Duncan
"A Sketch of the Customs, Manners and Way of Living of the Natives in the Barren Country about Nipigon." In Les Bourgeois de la Compagnie du Nord Ouest, ed. R.L. Masson, Vol. 2, pp. 297-98.

Campbell, Henry C.
Early Days on the Great Lakes: The Art of William Armstrong. McClelland and Stewart, Toronto, 1971.

Campbell, M.
The North West Fur Company. Macmillan, Toronto, 1957.
---. The North West Company. Rev. ed. Macmillan, Toronto, 1973.

Canada. Census.
Queen's Printer, Ottawa, 1851-1951.

Canada. Department of the Postmaster General.
Canada Official Postal Guide. Government Printing Bureau, Ottawa, 1910-1933.

Canada. Department of Railways and Canals.
Annual Report of the Department of Railways and Canals....Imprint varies, Ottawa, 1880-1937.
Regulations for the Dominion Canals, 1895. Government Printing Bureau, Ottawa, 1897.

Canada. Department of Transport. Dominion Bureau of Statistics.
Canal Statistics.... Imprint varies, Ottawa, 1925-45.

Canada. Parliament.
Debates of the House of Commons. Queen's Printer, Ottawa, 1867-1945.
---. Sessional Papers. Queen's Printer, Ottawa, 1867-1945.

Canada (Province). Department of Public Works.
"Keefer Report," Journals of the Legislative Assembly, 1852-53, Vol. 9, No. 3 (1853), Appendix Q: "Sault Ste. Marie Canal," Aug. 19, 1852. (unpaginated)

Canada (Province) Legislative Assembly
Journals.
Statutes

Canada. Public Archives, Manuscript Division.
RG3, Post Office.
RG8, C Series, British Military and Naval Records, Correspondence.
RG9, Militia and Defence.
RG10, Indian Affairs.
RG11, Public Works.
RG12, Transport.
RG43, Railways and Canals.
MG11, CO42, Vol. 70.
MG24, Notes on "a meeting of the Divisional Directors of the Victoria Mining Company," Aug. 5, 1856.
WO55, "Memorandum on the Fortifications at Kingston and on the Defence of the Frontier."

Canadian Pacific Railways Archives, Montreal
Photograph Collection.
Letterbooks.
C.P. Bridge Company at the Sault.

Carter-Edwards, Dennis
"The Sault Ste. Marie Canal." Research Bulletin No. 119, Parks Canada, Cornwall, 1980.

Carver, J.
Travels Through North America in the Years 1766, 1767, 1768. London, 1778.

Catlin, George
Letters and Notes on the Manners, Customs and Conditions of the North American Indians, 1832-1839. Dover, New York, 1973. Vol. 2.

Charlton Standard Catalogue of Canadian Paper Money
N.p., Toronto, 1980.

Clergue, F.
An Instance of Industrial Evolution in Northern Ontario, Dominion of Canada. N.p., 1900.

Constitutional (St. Catharines)
May 19, 1870.

Conway, Thor
Archaeology in Northeastern Ontario. Ministry of Culture and Recreation, Toronto, 1981.
---. The Point Louise Site: Report of the Regional Archaeologist. Ontario Ministry of Culture and Recreation, Sault Ste. Marie, 1975.

Coyne, J.H., ed.
"Galinee's Narrative." Ontario Historical Society Papers and Records, Vol. 4 (1903).

Creighton, D.
The Empire of the St. Lawrence. Macmillan of Canada, Toronto, 1956.

Cuthbertson, George C.
Freshwater: A History and a Narrative of the Great Lakes. Macmillan of Canada, Toronto, 1931.

Davidson, G.C.
The North West Company. Russell and Russell, N.Y., 1967.

Dobbs, Arthur
An Account of the Countries Adjoining to Hudson's Bay. J. Robinson, London, 1744.

Douglas, R. Allan, ed.
John Prince: A Collection of Documents. The Champlain Society, University of Toronto Press, Toronto, 1980.

Dunning, R.W.
Social and Economic Change Among the Northern Ojibwa. University of Toronto Press, Toronto, 1959.

Eldon, Donald
"The Career of Francis Hector Clergue." Explorations in Entrepreneurial History, Vol. 3 (April 1951). Boston.

Engineering Record
"The Accident at the Canadian Lock at Sault Ste. Marie." Vol. 59, No. 25, p. 790.

Ewing, Robert G.
"The Changing retail Structure of Sault Ste. Marie, Ontario, 1901-1971." PhD dissertation, University of Edinburgh, 1973.

Ferris, John, ed.
Fifty Years of Labour in Algoma: Essays on Aspects of Algoma's Working Class History. Algoma University College, Sault Ste. Marie, 1978.

Fowle, Otto
Sault Ste. Marie and its Great Waterway. G.P. Putnam's Sons, New York, 1925.

Franchère, Gabriel
Relation d'un Voyage à la côte du Nord Ouest de l'Amérique Septentrionale, dans les années 1810, 1811, 1812, 1813, 1814. C.B. Pasteur, Montreal, 1820.
---. Franchère's Journal. Champlain Society, Toronto, 1969.

Galbraith, J.S.
The Hudson's Bay Company as an Imperial Factor, 1821-1869. University of Toronto Press, Toronto, 1957.

Glazebrooke, G.P. de T.
A History of Transportation in Canada. McClelland and Stewart, Toronto, 1970. 2 vols.

Globe (Toronto)
1847-1870.

Grant, Peter
"The Saulteux Indians about 1804." In Les Bourgeois de la Compagnie du Nord Ouest. Ed. R.L. Masson. Reprint of 1889 ed. Antiquarian Press, New York, Vol. 2, 1960.

Griffin, J.B., ed.
Lake Superior Copper and the Indians. University of Michigan Press, Ann Arbor, 1961.

Harper, J.
Paul Kane's Frontier. University of Toronto Press, Toronto, 1971.

Harrison, R.
"The Break at the Canadian Canal." Inland Seas, Vol. 35 (Summer 1979), pp. 104-109.

Harvey, Chas. T.
"Pioneer Sault Canal." In Semi-Centennial Reminiscences of the Sault Canal. J.B. Savage, Cleveland, 1905.

Hatcher, Harlan, and E.A. Walter
A Pictorial History of the Great Lakes. Bonanza Books, New York, 1963.

Havighurst, Walter
The Long Ships Passing: The Story of the Great Lakes. Macmillan, New York, 1942.

Heisler, J.P.
"The Canals of Canada." Canadian Historic Sites: Occasional Papers in Archaeology and History, No. 8, 1973. Ottawa.

Henderson, George Fletcher
Federal Royal Commissions in Canada 1867-1966: A Checklist. Toronto, 1967.

Henry, Alexander
Travels and Adventures in Canada and the Indian Territories, 1760-1766. Hurtig, Edmonton, 1969.

Heriot, George
Travels Through the Canadas, 1806. R. Phillips, London, 1807.

Hickerson, H.
The Chippewa and Their Neighbours: A Study in Ethnohistory. Holt, Rinehart and Winston, New York, 1970.
---. The Southwestern Chippewa: An Ethnohistorical Study. Memoirs of the American Anthropological Association, No. 92, 1962.
---. "The Feast of the Dead Among the Seventeenth Century Algonkian Indians of the Upper Great Lakes." American Anthropology, Vol. 42 (1960).

Hornick, Leigh
The Call of Copper: A History of Bruce Mines and Area. North Shore Printing, Bruce Mines, 1969.

Horsman, Reginald
"British Indian Policy in the Northwest, 1807-1812." Mississippi Valley Historical Review, Vol. 45 (1958-59).

Hudson's Bay Company Records (Public Archives Canada, MG20)
Sault Ste. Marie Journals, 1824-1836, HBC microfilm reel 1m 131.
Sault Ste. Marie Correspondence, Copy Books, 1824-1853, HBC microfilm reel 1m 224, 245.
Sault Ste. Marie Correspondence, Inward, 1824-1861, HBC microfilm reel 1m 381.
Sault Ste. Marie Annual Reports, 1825-1835, HBC microfilm reel 1m 782.

Huron Signal (Goderich)
Nov. 4, 1852.

Illinois Historical Collection
"Hennepin's Narrative from La Louisiane." 1903.

Innis, H.A.
The Fur Trade. Rev. S.D. Clarke and W.T. Easterbrook. University of Toronto Press, Toronto, 1964.
---. The Fur Trade in Canada. University of Toronto, Toronto, 1962.

Jameson, Anna
Winter Studies and Summer Rambles in Canada. McClelland and Stewart, Toronto, 1965.

Jenness, D.
The Indians of Canada. 6th ed. National Museum of Canada, Ottawa, 1963. Bulletin No. 65, Anthropology Series No. 15.

Jones, Peter
History of the Ojibway Indians. W.A. Bennet, London, 1861.

Johnston, C.
Highways and Byways of the Great Lakes. Macmillan and Co., New York, 1911.

Johnston, J.
"An Account of Lake Superior." In Les Bourgeois de la Compagnie du Nord Ouest, ed. R.L. Masson, Antiquarian Press, New York, 1960, Vol. 2.

Judson, Clara Ingram
The Mighty Soo. Follet Publishing, New York, 1955.

Kehoe, J.J.
"The Sault Ste. Marie Ship Canal." Canadian Magazine, Vol. 1, 1893.

Kellog, L.P., ed.,
Early Narratives of the North West, 1634-1699. Scribner, New York, 1917.

Kingston, H.G.
Western Adventures or a Pleasure Tour in the Canadas. London, 1856.

Kinietz, W. Vernon
The Indians of the Western Great Lakes, 1615-1760. University of Michigan Press, Ann Arbor, 1940.

Legget, Robert
Canals of Canada. Douglas and Charles, Vancouver, 1976.

Leighfield, W.E.
"Some Phases of the Development of Algoma District and the City of Sault Ste. Marie." MA thesis, Queen's University, 1947.

Logan, W.E.
Report on Lake Superior Mining. Montreal, 1846.

Long, John
Voyages and Travels of an Indian Interpreter and Trader. London, 1791.

Macdonell, Allan
The North-West Transportation, Navigation and Railway Company: Its Objectives. Toronto, 1858.

MacGibbon, D.
The Canadian Grain Trade. Macmillan and Co., Toronto, 1932.

Malkus, A.
Blue-Water Boundary: Epic Highway of the Great Lakes and the St. Lawrence. Hastings House, New York, 1960.

Marcolin, Lorenzo
"Canadian Pacific Railway Company Steamship Lines — Last of an Era." Inland Seas, Vol. 22 (1966), pp. 3-16.

Masson, R.L., ed.
Les Bourgeois de la Compagnie du Nord Ouest. Reprint of 1889 ed. Antiquarian Press, New York, 1960. 2 vols.

McCracken, H.
George Catlin and the Old Frontier. Dial Press, New York, 1959.

McDonald, Graham A.
"The Ancient Fishery at Sault Ste. Marie." Canadian Geographic Journal (Dec. 1966), pp. 198-207.
---. "The Saulteur-Ojibwa Fishery at Sault Ste. Marie, 1640-1920." MA thesis, University of Waterloo, 1978.

McIlwraith, Thomas F.
"George Laidlaw." Dictionary of Canadian Biography, Vol. 11, pp. 481-83.

McLean, John
Notes on Twenty-Five Years Service in the Hudson's Bay Territory, 1849. Champlain Society, Toronto, 1932.

McPhedron, M.
Cargoes of the Great Lakes. Macmillan and Co., Toronto, 1956.

Mills, J.
Our Inland Seas, Their Shipping and Commerce for Three Centuries. McClurg, 1910.

Morris, Alexander
The Treaties of Canada with the Indians. Belford and Clarke, Toronto, 1880.

Morton, W.L.
Manitoba: A History. Toronto, 1957.

Nelligan, F.J.
"Catholic Missionary Labours on the Lake Superior Frontier, 1667-1751." Ontario Historical Society Papers and, Records, 51 (1959).

Neu, I.
"The Building of the Sault Canal." Mississippi Valley Historical Review, Vol. 40 (June 1953), pp. 25-46.

Niagara Chronicle
Jan. 2, 1853.

Nock, David
"E.F. Wilson: Early Years as a Missionary in Huron and Algoma." Journal of the Canadian Church Historical Society, Vol. 15, No. 4 (Dec. 1973), pp. 78-96.

Nock, O.D.
The Algoma Central Railway. Adam and Charles, London, 1975.

Nute, Grace Lee
Lake Superior. Bobs-Merrill, Indianapolis, 1944.
---. The Voyageur. Minnesota Historical Society, St. Paul, 1955.
---. "The American Fur Company's Fishing Enterprises on Lake Superior." Mississippi Valley Historical Review, Vol. 13, No. 4, (1926).
---. "The Papers of the American Fur Company: A Brief Estimate of Their Significance." American Historical Review, Vol. 23, 1927.

Parks Canada, Cornwall Office, Sault Ste. Marie Records
Canals: Construction, Maintenance and Repairs.
Dam, Weirs, Swing Dam: Construction, Maintenance and Repair Land Acquisitons.
Deeds: Algoma Central and Hudson Bay Railway Company Ontario, Hudson's Bay and Western Railway Company Pacific and Atlantic Railway Company.
Various correspondence with federal and provincial governments.

Parks Canada. Sault Ste. Marie.
Lock Books.
Lockmaster's Journal.
Report of the Arbitration Hearing into the Construction of the Sault Ste. Marie Canal. Ottawa, 1899
Superintendent's Annual Reports, 1896-1926.
Correspondence.
Maps, plans and drawings.
Photographs.
Operations of the Locks.
The Electrical Engineer.

Phillips, Paul C.
The Fur Trade. University of Oklahoma Press, Norman, Okla., 1961.

Prior, L.
"Sault Ste. Marie and the Algoma Steel Ltd." MA thesis, University of Toronto, Toronto, 1956.

Province of Ontario, Archives of Ontario
W. McTavish Letterbook (Hudson's Bay Co.).
Department of Crown Lands, Crown Lands Papers.
P. Russell Papers.

Punch, Katherine
"Sault Ste. Marie." Canadian Geographic Journal (Dec. 1966), pp. 198-207.

Pye, E.G.
Geology and Scenery. The North Shore of Lake Superior. Department of Mines, Toronto, 1969.

Quaife, M., ed.
The John Askin Papers. Detroit Public Library, Detroit, 1931.

Quimby, G.I.
Indian Life in the Upper Great Lakes, 11,000 B.C. to A.D. 1800. University of Chicago Press, Chicago, 1960.
---. Indian Culture and European Trade Goods: The Archaeology of the Historic Period in the Western Great Lakes Region. University of Wisconsin Press, Madison, Wis., 1966.

Rattray, W.J.
The Scot in British North America. Maclear and Co., Toronto, [1880-84] 4 vols.

Rich, E.E.
The Hudson's Bay Company, 1660-1870. Toronto, 1958-59. 3 vols.
---. Montreal and the Fur Trade. Beatty Memorial Lecture, McGill, Montreal, 1966.
---. Isham's Observations and Notes on Hudson's Bay, 1743-1749. Hudson's Bay Record Society, London, 1949.
---. Letters from Hudson's Bay. Hudson's Bay Record Society, London, 1972.

Riddell, W.R.
"The Last Official Report of the French Posts in the Northern Part of North America." Ontario Historical Society Papers and Records, Vol. 28 (1932), p. 134.

Ritzenhaler, R. and O.
The Woodland Indians of the Western Great Lakes. Natural History Press, Garden City, 1970.

Rogers, E.S.
Ojibwa Fisheries in Northwestern Ontario. Ontario Ministry of Natural Resources, Commercial Fish and Fur Branch, 1972.

Royce, C.C.
Cessions of Land by Indian Tribes to the United States: Illustrated by those in the State of Indiana. Annual Report of the Bureau of Ethnology. First Report, 1879-80. Washington, 1881.

Russell, N.V.
The British Regime in Michigan and the Old Northwest, 1760-1796. Carleton College, Northfield, Minn., 1939.

Sault Ste. Marie Historical Society Museum
Annual Reports.
Picture Collection.
Fur Trade File.

Sault Ste. Marie, Michigan, Bayliss Library
Picture Collection.

Sault Ste. Marie, Ontario, Public Library
The Sault Daily Star, 1852-1943.

Photo Albums of Canal Construction, 3 vols.
The Sault Ste. Marie and 49th Field Regiment Historical Society Museum.

Schoolcraft, Henry R.
Information Respecting the Conditions and Prospects of the Indian Tribes of the United States. Lippincote, Gramlo and Co., Philadelphia, 1852.
---. Narrative Journal of Travels from Detroit Northwest through the Great Chain of American Lakes in the Source of the Mississippi River in the Year 1820. Michigan State College Press, 1953.
---. Personal Memoirs of a Residence of Thirty Years with the Indians on the American Frontier with Brief Notes and Opinions. Arno Press, New York, 1975.
---. Narrative Journal of Travels Through the Northwestern Regions of the United States. E. Hosford, Albany, 1821.

Schull, Joseph
Ontario Since 1867. McClelland and Stewart, Toronto, 1978.

Sinclair, Duncan
"Exploration Line from Montreal River on the East to Michipicoten Harbour on the West." In Remarks on Upper Canada Surveys, Hunter and Rose, Ottawa, 1967.

Stanley, G.F.G.
"The First Indian Reserves in Canada." Revue d'histoire d'Amérique française, Vol. 4, No. 2 (1950-51).

Statutes of Ontario
"An Act to incorporate the Town of Sault Ste. Marie," April 23, 1887.

Sullivan, Alan
The Rapids. University of Toronto Press, Toronto, 1972.

Surtees, R.J.
"The Development of an Indian Reserve Policy in Canada." Ontario History, Vol. 61, No. 2 (1969).

Swainson, Donald
"The North-West Transportation Company: Personell and Attitudes." Historic and Scientific Society of Manitoba, Transactions, Series 3, No. 26 (1969-70).

Taylor, Gordon Garfield
"The Mississauga Indians of Eastern Ontario, 1634-1881." MA thesis, Queen's University, Kingston, 1981.

Thwaites, R.G., ed.
Jesuit Relations and Allied Documents. Burrows, Cleveland, 1896.

Turner, F.J.
The Character and Influence of the Indian Trade in Wisconsin: A Study of the Trading Post as an Institution, 1891. Benjamin Franklin, New York, 1970.

U.S. Commission of Fish and Fisheries
Report of the Commissioner for 1887. Government Printing Office, Washington, 1891. Part 15.

van Every, M.
"Francis Hector Clergue and the Rise of Sault Ste. Marie as an Industrial Centre." Ontario History, Vol. 56 (Sept. 1964).

Washington, U.S. National Archives
RG107, Registered Correspondence to the Secretary of War.

Warren, William
History of the Ojibwa Nation. Ross and Haines, Minneapolis, 1970.

Whitaker, R.
"Sault Ste. Marie Michigan and Ontario — A Comparative Study of Urban Geography." Bulletin of the Geographical Society of Philadelphia, Vol. 32 (1934).

Willoughby, W.R.
"The St. Mary's: Our First Ship Canal." Inland Seas, Vol. 11 (1955).

Wright, Nancy
Focus on Sault Ste. Marie: A List of Maps showing Sault Ste. Marie in Transition. Sault Ste. Marie Public Library, Sault Ste. Marie, 1975.

Young, A.E.
The Two Soos: American and Canadian. James Bayne, Grand Rapids, Mich., n.d.

Young, Anna
The Great Lakes Saga. Richardson, Bond, Wright, Owen Sound, 1965.